AF602070

Mitigating the Impact of Extreme Natural Events in Developing Countries

About the Centre

The Centre for Science and Technology of the Non-Aligned and Other Developing Countries (NAM S&T Centre) is an inter-governmental organisation with a membership of 47 countries spread over Asia, Africa, Middle East and Latin America. Besides this, 11 S&T agencies and academic/research institutions of Bolivia, Brazil, India, Nigeria and Turkey are the members of the S&T-Industry Network of the Centre. The Centre was set up in 1989 to promote South-South cooperation through mutually beneficial partnerships among scientists and technologists and scientific organisations in developing countries. It implements a variety of programmes including international workshops, meetings, roundtables, training courses and collaborative projects and brings out scientific publications, including a quarterly Newsletter. It is also implementing 8 Fellowship schemes, namely, Joint NAM S&T Centre – ZMT Bremen Fellowship; Joint NAM S&T Centre – ICCBS Fellowship; Joint NAM S&T Centre – DST (South Africa) Training Fellowship on Minerals Processing and Beneficiation; NAM S&T Centre – U2ACN2 Research Associateship in Nanosciences and Nanotechnology; Joint CSIR/CFTRI (Diamond Jubilee) – NAM S&T Centre Fellowship; NAM S&T Centre Research Fellowship; NAM S&T Centre – ACENTDFB Fellowship in Neglected Tropical Diseases and Forensic Biotechnology; NAM S&T Centre - ASRT, Egypt Fellowship Programme. These activities provide, among others, the opportunity for scientist-to-scientist contact and interaction, training and expert assistance, familiarising the scientific community on the latest developments and techniques in the subject areas, and identification of technologies for transfer between member countries. The Centre has so far brought out 80 publications and has organised 114 international workshops and training programmes.

For further details, please visit www.namstct.org or write to the Director General, NAM S&T Centre, Core 6A, 2nd Floor, India Habitat Centre, Lodhi Road, New Delhi-110003, India (Phone: +91-11-24645134/24644974; Fax: +91-11-24644973; E-mail: namstcentre@gmail.com; namstct@bol.net.in).

Mitigating the Impact of Extreme Natural Events in Developing Countries

— *Editors* —

R. J. Durrheim

South African Research Chair of Exploration, Earthquake and Mining Seismology
University of the Witwatersrand, Johannesburg, South Africa
Raymond.Durrheim@wits.ac.za

B. G. N. Sewwandi

Department of Zoology and Environmental Management
Faculty of Science, University of Kalaniya,
Kalaniya, Sri Lanka
sewwandidh@kln.ac.lk

CENTRE FOR SCIENCE & TECHNOLOGY OF THE NON-ALIGNED AND OTHER DEVELOPING COUNTRIES (NAM S&T CENTRE)

2019

DAYA PUBLISHING HOUSE®
A Division of
ASTRAL INTERNATIONAL PVT. LTD.
New Delhi – 110 002

ISBN: 9789389569179 (Int. Edition)

Published by : **Daya Publishing House®**
A Division of
Astral International Pvt. Ltd.
– ISO 9001:2008 Certified Company –
4736/23, Ansari Road, Darya Ganj
New Delhi-110 002
Ph. 011-43549197, 23278134
E-mail: info@astralint.com
Website: www.astralint.com

Printed at : **Replika Press Pvt. Ltd.**

Dr. Satish R. Wate
Former Director
CSIR-National Environmental Engineering Research Institute (NEERI),
Nagpur, India

And

Chairman
CSIR- Recruitment & Assessment Board (RAB)
New Delhi, India

Foreword

Extreme natural events are short-term changes in the weather or environment that can have long-term effects. An event occurring naturally becomes a natural disaster that has large-scale effects like earthquake, hurricane, avalanches, wildfires, flooding or drought, tornadoes, snowstorms etc. on environment and people.

Climate change is one of the most widely discussed issues in the present context and its role is more realized while understanding and analysing natural disaster. Climate changes have been occurring ever since the atmosphere was created on and those climate changes were influenced only by natural phenomena, prior to the origin of man. But in the modern era, the natural phenomena along with human activities are very strongly contributing towards the climate change, especially in warming up the globe. Climate change is the net result of many processes taking place on the earth and leading to the alterations in frequency, intensity, spatial extent and duration of weather and climate extremes.

Extreme natural events have long term impact on human, ecology and environment leading to heavy economic losses. Such losses become frequent in the past few decades, impacting everyone severely, especially developing countries. For undertaking the disaster prevention and risk reduction processes, understanding of the geological, meteorological, sociological, technical and economical aspects of the potential disaster is required. This becomes a challenging agenda for the nations to take up. Under such circumstances it is highly essential that nations come together to make collective efforts to educate, train and network individuals and organisations in searching science and technology based solutions to mitigate extreme natural disasters.

I am pleased NAM S&T Centre in partnership with the National Science & Technology Commission), Sri Lanka and the Research Centre-Technology for Disaster Prevention, South Eastern University of Sri Lanka (RC-TDP, SEUSL) organized an international Roundtable on **'Impacts of Extreme Natural Events: Science & Technology for Mitigation (IRENE)'** in Colombo, Sri Lanka in December 2017 to address issues related to prevention and mitigation of the impacts of extreme natural events amongst developing countries. This book, which is a follow-up of the workshop, comprises 24 scientific papers authored by experts from developing countries and address the impacts of natural extreme events in their respective countries and the mitigating efforts that have been initiated by their Governments. Several of these contributions focus on the solutions for prevention of natural calamities by adapting predictive assessment of previous events. This kind of roundtables can help the policy makers to do self assessment of the mitigating strategies in case of any natural disaster in future.

I compliment NAM S&T Centre for bringing out this valuable publication which will be of immense use to policy makers, researchers, civil societies and all others engaged in disaster mitigation related issues.

(Satish R. Wate)

Preface

The human, economic and environmental impacts of extreme natural events (such as floods, cyclones, earthquakes and tsunamis) are often greatest in developing countries because of their large populations and limited capacity to implement mitigation measures. The frequency and intensity of some of these events will likely be exacerbated by climate change. The importance of disaster risk reduction is embedded in many of the Sustainable Development Goals adopted by the United Nations General Assembly in 2015 and is a focus area of the International Science Council. However, the solutions developed by wealthy and technologically-advanced countries cannot always be successfully applied in developing countries. Thus it is important that scientists, engineers and disaster managers from the developing world share their experience and expertise to improve the life chances of the most vulnerable communities.

This volume contains papers presented at the International Roundtable on '*Impact of Extreme Natural Events: Science and Technology for Mitigation* (***IRENE***)' in Colombo, Sri Lanka from 13-15 December 2017. The papers provide examples of research and development of impact-mitigating technologies and strategies in developing countries. It is hoped that these examples will stimulate innovation and implementation by natural and social scientists and engineers, and will assist decision-makers to develop and apply evidence-based policies and procedures.

The first two sections of the volume contain papers that address the risks posed by hydro-meteorological hazards, such as lightning strikes, floods and droughts. It continues with a section devoted to the impacts of climate change. Next are two sections that deal with geo-hazards. Landslides are phenomena that may be triggered either by excessive rain or by earthquake shaking. The earthquake section

contains a single paper that describes the assessment of seismic hazard assessment along the East African Rift System. The final section of the volume contains papers that address disaster risk reduction from a multi-hazard or integrated perspective.

The volume concludes with the Colombo Resolution on 'Mitigation of Impact's of Human Hazards due to Extreme Natural Events: Five Year Road-Map on the Adoption and Development of Scientific and Technological Advancements', adopted by all participants in the International Roundtable.

Section I: Lightning Strikes

This section consists of papers that discuss the impact caused by lightning and measures to minimize injury and damage. ***Chapter-1*** provides an overview of the impact of lightning strikes in India and measures to mitigate injury and damage, while ***Chapter-2*** focuses on the situation in the Odisha State, Eastern India.

Section II: Flood, Drought and Water Pollution

This section consists of nine papers on flood, drought and ground water pollution in India, Sri Lanka, Togo, Vietnam and Zambia. ***Chapter-3, 4*** and ***5*** provide case studies from Sri Lanka. ***Chapter-3*** discusses the effects of flood on livestock in the Ampara District of Sri Lanka. It provides information on impact of floods on livestock and the mitigation measures implemented by the government and non-government organizations. Technological and non-technological measures to minimize the effect of floods on livestock are proposed. ***Chapter-4*** focuses on the role that flooding plays in triggering outbreaks of outbreaks of dengue fever, while ***Chapter-5*** shows how GIS can be used to mitigate the risk of flooding. ***Chapters-6*** and ***7*** apply GIS and machine-learning technologies to the assessment and mitigation of flooding in Togo and Vietnam, respectively. Land use dynamics in a wetland in India are assessed in ***Chapter-8***, which shows that wetland encroachment will increase the floods. In ***Chapter-9***, the focus shifts from floods to drought, showing how MODIS satellite remote sensing and GIS is applied in Zambia. ***Chapter-10*** is a review study on the impact of both floods and droughts on agricultural production in the Batticaloa District in Sri Lanka. It also gives details of the projects implemented to reduce the risk and has recommended technological solutions to mitigate impacts on agricultural production.

Section III: Climate Change

This section provides information on the rate of climate change, the expected impact of climate change on socio-economic sectors, and mitigation measures that have been applied in Afghanistan, India, Iran, Malawi and Sri Lanka. ***Chapter-11*** describes a study to identify the impact of climate change on Afghanistan. It found that public knowledge is minimal, and the researchers propose strategies to increase awareness. ***Chapter-12*** presents a study of the impact of climate change on economic sectors in the rural areas of Karnataka, India. Further, the strategies to be implemented to cope with these effects are well explained by the authors.

Chapter-13 assesses a wide range of approaches to mitigate the impacts of climate change in India. These include desalination, interlinking or diversion of rivers, cloud ionization, development of deep aquifers and the recycling of wastewater. ***Chapter-14*** describes efforts to use indigenous knowledge of plants, insects and birds to ensure food security in Malawi; for example, by identifying the optimum time to plant crops. Climate change and extreme weather will also contribute to changes in air quality, with an inevitable impact on human health. In ***Chapter-15*** reviews recent work on the links between these phenomena in Sri Lanka. Quantification of the impact caused by extreme natural events is essential in successfully implementing mitigation measures. One such quantification method used in Iran to assess desertification is given in ***Chapter-16***, which gives the details on how to analyse the desertification and the factors to be considered in analysing.

Section IV: Landslides

This section presents two papers related to landslides. Assessment and monitoring of the impact of landslides is essential to safeguard lives and protect properties. ***Chapter-17*** discusses the techniques for assessment and monitoring of landslides in Indian Himalayas. 'Soil nails' are a technology that may be used to stabilise landslide-prone slopes. ***Chapter-18*** describes laboratory tests and numerical simulations factors that affect their performance.

Section V: Earthquakes

Earthquakes pose a significant risk in the developing world, as structures are rarely earthquake-resistant. The East African Rift System is a seismically-active region that stretches from Ethiopia in the north to Mozambique in the south, and passes through Kenya, Uganda, Tanzania, the Democratic Republic of Congo, Rwanda, Burundi and Malawi. ***Chapter-19*** assesses the seismic hazard in these countries, which can be used to formulate appropriate building codes.

Section VI: Integrated Disaster Risk Reduction

Many developing countries are exposed to several natural hazards; thus strategies to reduce the risk of disasters should be considered holistically. Capacity-building of people at all levels is important in disaster risk reduction. ***Chapter-20*** provides a Myanmar experience on capacity building and strategies for successful implementation of science and technology for mitigating impact of natural extreme events. The increased risk of extreme natural events shows the importance and the need of moving from planning aspects to implementing strategies to reduce the impact of such events. Review on a study carried out on landscape restoration techniques is given in ***Chapter-21*** along with its applicability to Indian context. Disaster notification system can reduce the risk of impact by natural disasters. A specific disaster notification system developed in Myanmar which sends the information only to the people living in the imminent disaster area given in ***Chapter-22***. In ***Chapter-23*** a number of Artificial Intelligence (AI) technologies that extract information from social media to aid disaster response and management are presented by researchers from the Qatar Computing Research Institute. ***Chapter-24*** provides a review on key issues associated with different strategies in mitigating risk from natural extreme events. Further, the authors have found the key issues and seek for solutions to develop a new approach for natural risk reduction and climate change adaptation.

R.J. Durrheim

B.G.N. Sewwandi

Introduction

The climate change at global level is leading to the alterations in weather and climate extremes. These climate extremes impact severely on both human and ecosystems including economic losses, sectors such as tourism and agriculture, urban settlements, small island states, *etc.* Scientists predict that the frequency and intensity of these disasters are likely to increase as a result of the effects of climate change. Their susceptibility is principally based on the geographical, geological and socio-economic characteristics. During the last few decades many countries in Asia and Africa encountered unforeseen natural disasters apparently due to both the climate change and deficiencies of the built environment. It seems that in the last five years such calamities have been accelerated such as Monsoon flooding in Bangladesh; Hurricane Irma in USA and Caribbean; floods in different regions of India, mudslide in Colombia and earthquakes in Mexico and Iran. These high-profile mega-disasters are raising global awareness of the need to build the capacity of national governments, civil society organizations and international entities to prevent, respond to and recover from natural disasters. Actions at multiple levels engaging different actors' *viz.*, national and international stakeholders and private institutions is the need of the hour to push the international protection regime forward.

Keeping this in view, the NAM S&T Centre in partnership with the National Science and Technology Commission (NASTEC), Sri Lanka and the Technology for Disaster Prevention, South Eastern University of Sri Lanka (RC-TDP, SEUSL) organized an international round table on '***Impact's of Extreme Natural Events: Science and Technology for Mitigation* (IRENE)**' in Colombo, Sri Lanka during 13-15 December 2017, which brought the scientists, experts and professionals engaged in R&D, policy making and implementation, social activists and other stake holders to a common forum for sharing views and experiences for the development of a road map for reducing the risks in real situations.

The round table was attended by 44 senior professionals from 18 countries – *Egypt, India, Indonesia, Iran, Iraq, Malaysia, Mauritius, Myanmar, Nepal, Pakistan, Palestine, Qatar, South Africa, Togo, United Kingdom, Vietnam* and *Zambia*, and the host country Sri Lanka.

As a follow up of Sri Lanka International round table, the present book **'Mitigating the Impact of Extreme Natural Events in Developing Countries'** has been edited to compile the contributions made by the participants. There are 24 scientific and technical papers contributed by the expert/participants from 12 countries.

I appreciate efforts put in place by both of the editors for technical editing of the manuscripts. I also acknowledge the interest and valuable efforts of the entire team of the NAM S&T Centre and am especially thankful to Dr. (Mrs.) Kavita Mehra, Dr. Vijay Verma, Mr. M. Bandyopadhyay, Ms. Rashmi Srivastava, Ms. Gloria Susan Cherian and Mr. Pankaj Buttan in compiling and checking the manuscripts, liaising with the authors, cover page designing, proof reading, formatting and taking all the necessary actions in giving a shape to this volume.

I am sure this book will be of immense use to all those associated with issues related to mitigation of mega disasters due to extreme natural events, from researchers to policy makers, non government organisations and government officials in the developing and other countries.

Dr. Amitava Bandopadhyay
Director General
NAM S&T Centre

Contents

Section III: Climate Change

Section I

Lightning Strikes

Chapter 1

Lightning: An Extreme Natural Event Causing Loss of Life and Property

R. Arora

Department of Electrical Engineering, Indian Institute of Technology, Kanpur – 208 016, Uttar Pradesh, India
E-mail: ravindra.arora@gmail.com

ABSTRACT

Apart from being a captivating phenomenon in nature, the fury and the uncontrolled fierceness caused by lightning could be highly dangerous to life, property and other assets on this earth. As per estimation made by 'The Federal Centres for Disease Control and Prevention', nearly one hundred people are killed in United States by Lightning each year. Unfortunately, this figure is much higher in India. According to 'National Crime Record Bureau of India', in the years 2013 and 2014, 2833 and 2582 people got killed, respectively due to lightning in India. As reported by 'Economic and Political Weekly', it is estimated that out of the total number of deaths between 1967 and 2012 due to extreme natural events like; floods, landslides, earthquakes, heat and cold waves and lightning in India, 39 per cent people died due to lightning alone, maximum number of victims of all natural calamities. Hence it can be concluded that the most severe impact of extreme natural calamity is caused by lightning in India killing maximum number of people. Besides the loss of human life, the lightning also causes destruction of property due to blasts and fire. A very large number of unaccounted cattle, both pets and wild are also killed. Although not so frequent in India, lightning also causes fire in the jungles, which becomes uncontrollable at times. An effort to reduce the damage caused by lightning is therefore urgently required, especially to protect the life of the innocents.

Keywords: *Lightning, Lightning safety, Lightning strikes, Lightning flash activity, Electrocution.*

1. Introduction

Our earth is not the only planet in this universe having lightning activity. In the year 2017 NASA reported occurrence of lightning on Venus. The phenomenon of lightning must have existed on our Earth even before the human life began.

The lightning activity serves to maintain a balance in the global electrical charge system. The charge produced in the atmosphere is enabled to get absorbed by the ground with the help of lightning phenomenon. The dissipation process of disposal of the charge injected by lightning strikes towards the inner ground is peaceful if it gets conducted or to say that it finds the least resistance path. However, unable to go in to the ground this charge may result in dangerous actions. The living being themselves and their property has become most vulnerable for the destruction caused by lightning. The damage caused by lightning strikes on this earth may not appear to be significant to the common people. But the statistical records show that it is quite substantial. This could be because the damage caused by lightning is isolated but well spread all over the earth surface, unlike the damage caused by other natural disasters, for example, earthquakes, tsunami, floods *etc.*

The combined effect of deforestation, industrialisation and effluent gas emission has led to global warming. It has influenced drastic change in climatic conditions and also the rain patterns all over the world. There is more of flash rain occurrence accompanied with vigorous thunder and lightning activity than continuous and sustained rains, which used to take place half a century ago. Those conditions used to result in much lesser lightning activity. The change in climatic conditions has affected the order of lightning and the damage caused by lightning strike activity tremendously. It is interesting to learn that the nature of damage caused by lightning differs from country to country. In United States the damage is more to the property. According to 'National Fire Protection Association' (NFPA) the local U.S. fire departments responded to an average of 22,600 fires started by lightning every year from 2007 to 2011 causing an average of only nine civilian deaths and loss/damage of direct property worth $ 451 million per year. Contrary to the most recent estimation of on an average only 31 people killed by lightning every year in U.S., 1755 people are killed every year in India. One has to go into the depth of the problem why such a huge loss of life is caused in the third world countries, whereas more of property damage takes place in advance country like United States of America due to lightning?

The situation demands a detailed analysis of the problem in view of the local conditions, educational level and the mind-set of the people, awareness and the technological development to be able to suggest large scale disaster management programmes/policies for the third world countries. Most simple and affordable methods need to be adopted at large scale in our countries to mitigate and minimise the effects of the natural calamity caused by lightning. This paper makes a thorough analysis and suggests most simple and affordable methods to reach the masses within the existing infrastructure for the third world for mitigation. Effective mitigation efforts of a natural calamity require knowledge, awareness and understanding of the severity of the problem. Making long term efforts on actions to reduce the extent of probable damage comes next.

2. Global Trends in Damage Caused by Lightning

Lightning is a continuous and continued activity all over the world but it is not uniformly distributed in different parts. It is especially high in the countries falling near the equator. About 70 per cent of lightning on land occurs in Tropics, where the majority of the thunder storms take place (Matt 2015). Satellite measurements revealed that at any given instant of time the total number of estimated lightning flash activity on our globe is of the order of (44 ± 5) flashes per second (Oliver 2005). It tracks the instant lightning activity around the globe continuously as shown in **Figure 1.1**. However, this system tracks the total lightning activity as it is unable to distinguish between the cloud to cloud and cloud to ground flashes. Only about 10 per cent of the total lightning flashes strike on the earth and out of these, not all result in causing damage. Still the total damage caused is quite significant. Lightning could take place at any time anywhere. However, the lightning activity shifts with the weather and its changing pattern on the globe. For example, with the change in climate from northern to southern hemisphere, the intensity of lightning activity also shifts.

The damage caused by lightning is related to the regional trends in development activities on the earth, especially the construction activity. There is certainly much more metal, the conductive material, on the surface of the world today than it used

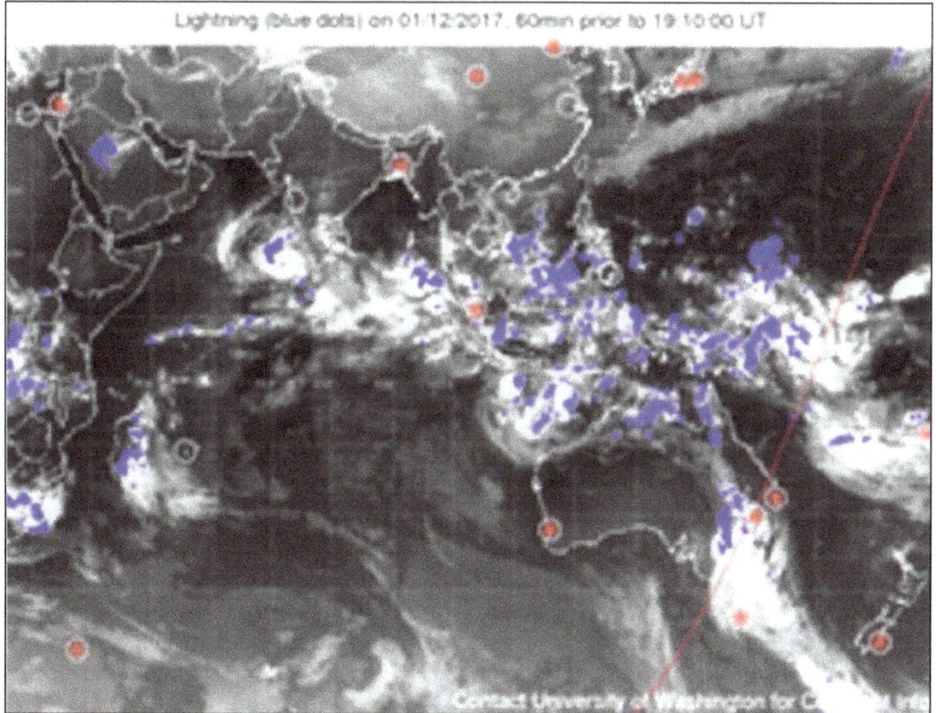

Figure 1.1: Distribution of Lightning Activity in South East Asia Region of the World on 2nd Dec 2017.

to be in the past. It is a well-known fact that the lightning is attracted towards conductors to strike because there is an enhancement of electric field at sharp metallic terminations in the air. The construction of high rise buildings, electrical power network, bridges, communication towers, rail and other transportation system networks *etc.* is ever increasing. The trends in residential and low rise warehouse structures *etc.* also determine the extent of regional damage. Hence, different patterns of damage caused by lightning are reported in different parts of the world.

The second important observation with regard to the damage caused by lightning is that the density of population in a particular region and the awareness and economic conditions of its inhabitants affects the extent of damage. The statistical data records reveal that the human life in the regions economically poorer and having high density of population is more endangered by lightning strikes than the property. A revealing comparison of the damage caused by lightning in USA and in India is made below.

3. Damages Due to Lightning

There are three major ways by which the lightning causes wide scale damage; fire, industry losses and life. The extent of the damage caused varies considerably from region to region and depend upon the local conditions. These may be density of population, economic conditions, awareness and trends. However, the losses caused by lightning can be categorised broadly in terms of the value of damage to the property in money and the loss of living being, which is invaluable.

3.1 Loss of Property

According to the 'US Department of Commerce' report published in the year 2014, although, the property damage by lightning is around US $48 million every year, there appears to be a conflict. The insurance industry report from 'Hartford Insurance Group', (www.tmcnet.com) released in the year 2006 says in contrast that the lightning is responsible for more than US $5 billion in total insurance industry losses annually.

A report compiled by Richard Kithil of '***The National Lightning Safety Institute*** (NLSI)' available at www.lightningsafety.com suggests that the total lightning related cost and losses may exceed to be much higher, of the order of US $ 8-10 billion per year in USA. The major cause of damage to the property include-

Fire

In the time period of year 2007 to 2011, an average of 22,600 fires started by lightning every year in USA according to a report published by 'National Fire Protection Association' in the year 2012. Out of these, lightning related structure fires alone were 58 per cent. The most vulnerable structure has been the church contributing 30 per cent of the total fires and the second most unprotected structure in US is the residential building, which had 18 per cent of the total fires. There were 7164 cases of '***forest fires***' reported in the year 2010 initiated by lightning strikes according to '*National Interagency Fire Centre*' USA, which is a tremendous loss. The loss of property estimates in South Africa is also quite high. The cost incurred by the government on insurance claims due to lightning in South Africa amounts to more than Rand 500 million per year (Gijben 2012).

Electrical Power Outages

According to Ralph Bernstein of '*Electric Power Research Institute (EPRI)*', annual average of 30 per cent of all the power outages were lightning related amounting to a total cost approaching to US$ 1 billion. Such an estimate of the value of loss of property due to lightning in India is not available. It appears to be a rare documentation in the third world countries, if at all it is available.

3.2 Loss of Life

The US "National Weather Service", run by 'National Oceanic and Atmospheric Administration' (NOAA) is hosting a very informative and educative site; www.lightningsafety.noaa.gov/victims.htm. According to their statistical data compilation, in US the human deaths due to lightning in the year 2017 till October, have been only 15. The last ten year average, from 2007 to 2016, is 31 fatalities. There has been a dramatic decline in lightning deaths since 1940 when it used to be around 400 every year. This is attributed to the decline in rural population, which decreased from 60 per cent in 1900 to about 25 per cent in 1990 (Lo'pez and Holle 1998). There has been a continued decline thereafter too, these figures are remarkably low when compared with the figures in India.

According to the '***National Crime Records Bureau***' in the year 2014 and in 2013, 2,582 and 2,833 people respectively got killed due to lightning strikes in India (Thiruppugazh 2018). As reported by Economic and Political Weekly, it is estimated that out of total number of deaths between 1967 and 2012 due to natural calamities like; floods, landslides, earthquakes, heat and cold waves and lightning, 39 per cent people died due to lightning alone (Illiyas *et al.*, 2014). This is the highest victim number by any single natural calamity in India. Hence, it can be concluded that in India the most damaging natural calamity is caused by lightning strikes killing maximum number of human being. Loss of animals and cattle wealth is not accounted for anywhere. The author is aware of the incidence in 2009 in Assam, India where a herd of about half a dozen wild elephants got electrocuted by direct lightning strike.

Accidents are often caused/reported where buildings under construction collapse due to lightning strike causing large number of deaths and also property damage as shown by following examples;

- A 1200 MW thermal power plant chimney under construction in Korba district, Chhattisgarh, India collapsed during heavy rain and thunder on 24th Sep, 2009 killing 20 people on the spot and trapping 50 (Khan 2012). The tall chimney at the BALCO plant was proposed to be built to a total height of 275 m, which was already constructed more than half of its proposed height. According to eye witness accounts, the huge structure came crashing down after being struck by lightning.
- In Sept. 2015 at Mecca, Saudi Arabia, lightning struck on a huge construction crane in a square where about 12,500 people were present,

killing 107 people and injuring hundreds. There were a number of similar cranes for construction present in that premises.

- When severe thunder storms and torrential rains lashed across Northern and Eastern India, Bihar, UP, MP and Jharkhand, 117 people were killed by lightning strikes just in two days, 21st and 22nd June, 2016 (Times of India, 23rd June, 2016).

It is unfortunate that loss of life and damage to property due to lightning in India are outside the purview of disaster related relief. It is not compensated by the 'Natural Calamity and Relief Funds' in India (Illiyas *et al.*, 2014).

4. Discussion on the Diversity in Damage Patterns

It is a fact of life that the dominating common nature of damage caused by lightning differs from country to country, region to region. It appears to be related with the local conditions, climate and life style, educational, cultural and technological development of the people living in the region.

In an advance country like US, in spite of high density of lightning strike, the loss of human life could be minimised over the last few decades (Mattie 2013). This could be because of high level of awareness induced among the general people. A good educational and information system and higher economy of the country do contribute towards awareness of the people. A systematic approach and efforts towards the mitigation of the problem faced by the people does bring results. However, on the other hand, the extremely high loss of property could be attributed to the casual approach towards resources, which determine the widely adopted trends in construction and design vulnerable to ignition of fire. Most of the residential buildings are constructed on wooden structure, which would catch fire easily by lightning strike. It is very unfortunate that on these structures even the simplest lightning protection scheme is not made mandatory. On one side the life is well protected through the excellent awareness programmes but on the other, a very casual approach is given towards providing protection to their wooden structures. It is a difficult trend to be understood.

The situation in a country like India and in the third world is entirely different. The masses are deprived. The educational and the economic background is so low that the people are unaware of the ways of protecting themselves under such situation. Deprived of information and guidance at the right time, the people are unable to protect themselves. It results in much higher level of avoidable calamity of loss of life. The level of awareness in India is such that in spite of the availability of technical knowledge and know how, even simple inexpensive techniques of lightning protection are rarely adopted. Only some high rise buildings and very few residential houses are provided with lightning protection.

The structural materials and the construction techniques in India are entirely different to those followed in US. Unlike the wooden structures, more of bricks and cement concrete are used for the construction of buildings. When such a structure having no lightning protection is struck by lightning, the charge injected being restless, circulates on the floors and wall surfaces in search of the least resistance path to conduct itself. On coming across a metallic object, which may be normally floating, the charge is injected back in to the atmospheric air, an insulating material. This process expands the air very fast, like an explosive does, creating a blast. Such a phenomenon is observed quite commonly when the cement concrete building structures are struck by lightning. Depending upon the local conditions and the magnitude of charge injected by the lightning stroke, the blast could shatter the structure itself, break glass window panes, lift objects and cause multiple kinds of damage. It may even ignite fires or char isolated insulating materials like paper, plastics, fine brush hair *etc.* on its way. Big fires and major injuries to living being are rare in such cases.

The collapse of the power plant chimney under construction, mentioned above, was due to such a blast created by lightning strike. When the charge injected could not find passage to go to ground in the absence of proper continuous 'lightning conductor', it created a blast. A clear case of unawareness, lack of technical training accompanied with gross negligence.

5. Mechanisms for Mitigation of Devastating Effects of the Extreme Event of Lightning

The analysis of available records reveals that there are three major types of damage caused by the extreme natural event of lightning. These are; loss of life, property, fires and blasts resulting in loss of life and property both. Depending upon the local conditions, the damage caused due to this natural calamity differs from region to region over the seasons on the globe. In view of the above, following simple measures are proposed to mitigate the adverse effects of lightning strikes.

5.1 Protection of Life

It is remarkable that the loss of life figures in US have come down drastically in the last century. Initially the first cause was decline in rural population over the years. But during the last a few decades, the credit goes to their initiatives and efforts for spreading the awareness among the people at large. Their well-educated society is made more conscious and aware with the help of educative websites, excellent coverage by media and the weather agency forecast systems. In US there are exclusive weather channels on the TV network available round the clock. These are devoted only to comprehensive weather reports, forecast, and educative precautions and warnings in case of emergent situations. The class 5th science curriculum in the schools includes a full chapter on lightning, which covers the basic knowledge of the phenomenon, its adverse effects and advises the precautions to be taken if caught under thunder. They have a popular slogan taught to the children; "***When Thunder Roars, Go Indoors***". First a few minutes are crucial for a person hit by lightning strike as it may even stop the heart beating, it is part of the training programme for security personnel and for the people in general to provide the First Aid to the victim. For us in the third world countries, the successful model of spreading the awareness in US is an ideal example.

5.2 Protection from Fires and Blasts on Structures

The fires and blasts caused by lightning on residential buildings not only result in huge loss in terms of money but also result in loss of human lives. Such incidents also induct a kind of phobia among the immediate affected people, a misery for large number of residents.

It is a common feature, practically all over the world, that the low rise buildings are not provided with any protection from lightning. The wooden structure buildings are constructed with sharp edges and corners on their slanted roofs making them vulnerable for lightning strikes. If an effective protective system, which could provide a low resistance path to the charge injected by lightning strike towards the ground, the chances of wood catching fire reduce considerably. The cement concrete and brick finish buildings besides having sharp edges and corners on their parapets also have often sharp and projecting metallic objects on their top most terrace, for example, tin-roofs, vertically projecting water pipes, metallic protective fence with sharp edges *etc.* These are generally electrically floating metallic bodies as they are not grounded. It is a well-known fact that these attract lightning to strike. Finding no conductive path to go inside the ground, the charge gets trapped and circulates restlessly on the building walls and floors.

All these structures must be provided with some simple and inexpensive lightning protection schemes in order to convince and encourage the owners to protect their buildings from eventual lightning strike. Although high-tech lightning protection gadgets, for example, electrically active '***Early Streamer Emission***' systems have been developed and are available commercially, the conventional system of installation of passive '***lightning conductor***' is still considered to be quite safe, simple and inexpensive method for lightning protection. As per various specifications, installation of simple lightning conductor protection schemes are recommended for different types and sizes of structures. It comprises in three parts, as shown in **Figure 1.2**.

The horizontal air terminal is recommended to run all along the edges, corners and parapets as these are most vulnerable spots where the lightning may strike. Details of recommended protection system for different types of structures are given in all national and international specifications on lightning protection.

A still simpler method is proposed for such buildings to provide protection from lightning. The cement concrete residential buildings install overhead water tanks in open, normally on their highest terrace level, which is vulnerable to lightning strikes. The overhead tanks are provided with water inlet as well as outlet pipes made of galvanised iron. Whereas the water inlet pipe is a continuation of the large scale underground water supply network, the outlet pipes from the overhead tanks terminate at various locations within the building, forming an electrically floating conductor. It has been observed that on such residential buildings if the lightning strikes, it strikes on the water pipe termination network on the terrace. A simple '**metallic bond**' between the pipes of this network would divert the charge inducted by lightning strike, on any such pipe, towards the lower resistance path provided by the underground network of the water supply pipes and get easily dissipated

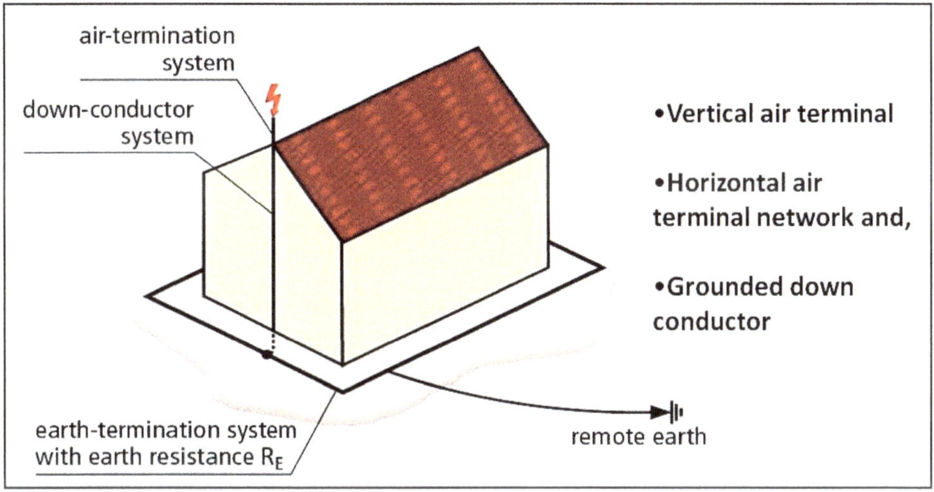

Figure 1.2: A Simple Lightning Conductor Protection System for Buildings.

inside the ground. Even if the water inlet is by a pumping system from a bore-well, such a provision would be able to provide low resistance path to the impulse current injected by lightning. Such an arrangement provides all the three parts of the lightning conductor protection system within, that is, the vertical air terminal, the horizontal air terminal and the grounded down conductor. Appropriate condition for such a provision may not be available with every construction, but with most of them. It does not cost anything, but just a skill.

6. Summary

The statistical data reveals that out of all the natural calamities of extreme natural events, highest human death toll is caused every year due to the single event of lightning in India. But it is unfortunate that loss of life and damage to property due to lightning in India are outside the purview of disaster related relief and hence there is no provision for providing a compensation to the victims by the '***Natural Calamity and Relief Funds***' organisation. The high death toll in India and probably in the third world countries due to lightning is attributed to; population density, lack of education, lack of awareness, provision of very poor weather forecast, warning, and precaution advisory system and last but not the least the ignorance of the local people of the region. Following measures are strongly recommended for mitigation:

- ✰ Inclusion of a brief chapter on lightning in the social science books of class 5 in the primary schools.
- ✰ An exclusive 'weather channel' on the TV network providing detailed weather report, forecast, information on severe conditions and safety measures/advise in order to mitigate the adverse severe effects.
- ✰ A comprehensive Website all on lightning related issues, including personal as well as property safety and simple methods of protection, owned by government or private sector.

In some parts of the world huge loss of property is inducted due to fires ignited by the lightning strikes on wooden structure buildings and in other parts the blast created by lightning strikes not only shatters the structure but also kills people. For mitigating this catastrophic effect the only way is through popularising the installation of simple and inexpensive lightning conductor protection system. It should be made mandatory for all the fire prone structures and on the tall structures even during their construction.

References

1. Gijben M. 2012. The Lightning Climatology of South Africa. *South African Journal of Science* 108: 3-4
2. Illiyas F.T., Mohan K., Mani S.K., Pradeep Kumar, A.P. 2014. Lightning Risk in India: Challenges in Disaster Compensation. *Economic and Political Weekly* 49: 23-27
3. Khan S. 2012. Chhattisgarh: Brazen Balco flouting rules in power plant construction. India Today, 2012.

   https://www.indiatoday.in/india/north/story/brazen-balco-flouting-rules-in-power-plant-construction-99891-2012-04-22
4. Lo'pez R. and Ronald L.H. 1998. Change in the Number of Lightning Deaths in the United States during the Twentieth Century. *Journal of Climate* 11: 2070-2077.

   (Available on journals.ametsoc.org)
5. Matt W. 2015. What causes lightning? *Universe Today* (July 10) https://phys.org/news/2015-07-lightning.html
6. Mattie Quinn 2013. Death by lightning a danger in developing countries. National Geographic.

   https://news.nationalgeographic.com/news/2013/11/131102-lightning-deaths-developing-countries-storms/
7. Oliver, J.E. 2005. Encyclopaedia of World Climatology. Springer Science+Business Media, Netherlands, Springer. ISBN: 978-1-4020-4870-8
8. Thiruppugazh V. Guidelines for Thunderstorm and Lightning/Squall/Dust/hailstorm and Strong wind. National Disaster Management Authority, August 2018.

   https://ndma.gov.in/images/pdf/Draft-Guidelines-thunderstorm.pdf

Chapter 2

Risk of Lightning and Mitigating its Impact in the Eastern Indian State of Odisha

B.C. Panda

Department of Civil Engineering,
Indira Gandhi Institute of Technology, Odisha, India
E-mail: bikashapanda@yahoo.com

ABSTRACT

Lightning is a sudden electrostatic discharge that occurs between electrically charged regions of a cloud and the ground during a thunderstorm. It generates extraordinary temperature and current causing loss of life and damage to property on earth in a strike. Lightning causes more deaths in India than any other natural disaster like by flood, landslide, heat stroke and cold waves. The case is severe for the eastern Indian state of Odisha, where on an average; more than 327 persons die of lightning strikes every year. It seems nationwide; less serious efforts have been done to study thoroughly the vulnerability of areas to lightning and its mitigation. Therefore, this research study will present the compilation of incidents of lightning fatality in India as well as Odisha so as to assess the gravity of the risks involved. Existing policy matters related to lighting hazard vis-à-vis natural calamities will be discussed. Mitigating measures to be adopted will be presented and efforts necessary for the region will be suggested. The methodology followed in this study is based on the data collected from field survey and various sources like district administration, Panchayat level, hospital, police station, disaster mitigation authority, relief commissioner office, newspaper, web sites and journals. The corresponding results have been presented with respect to month, district, age group, and instant location etc. The conclusion is that the government shall be more serious about lightning fatalities. Early warning system and mitigating efforts shall be deployed in vulnerable areas along with educating the mass through awareness program.

Keywords: *Lightning risk, Indian condition, Odisha, Field survey, Mitigation, Zonal mapping, Awareness.*

1. Introduction

Lightning is a high-energy luminous electrical discharge from a cumulonimbus cloud to the ground through a conductive path accompanied by thunder causing losses to life and damage to properties. The peak discharge current varies from many thousand amperes to 2,00,000 amperes or more and its passage is severely damaging to humans, trees, electrical structures, home appliances, and other living and non-living objects (IEEE, 2005). Most direct damage results from the heavy return stroke current that produces a large temperature rise to above 30,000 K (29,726 °C) in the resistance of the channel through which the charge travels (Rafferty, 2011). The heating of air leads to a sudden pressure increase and production of a powerful shock wave that is heard as thunder. Always, the current forms arcing at the point of attachment and when it takes place in a combustible or explosive environment, fire or an explosion can result. If the lightning current is carried by an enclosed conductor (*e.g.*, within a jacketed cable, through a concrete wall, or beneath a painted surface), entrapped moisture is turned into high pressure steam, which can cause the cable or painted object to burst, the wall or a tree to explode, or the shoes to be blown off the damp feet of a person struck by lightning. Lightning can pose a threat to numerous processes associated with the production and handling of energetic materials, nuclear stockpiles and materials that may produce environments containing flammable gases, flammable or combustible liquid-produced vapors, combustible dusts, or ignitable fibers/flying (Guthrie and Rousseau, 2014). Often, failure of internal systems takes place due to lightning electromagnetic impulse in telephone exchanges, computer systems, and controllers of various establishments. Finally, loss of human life, loss of services to public and loss of economic values and heritage structures may result from damage due to lightning (BS EN/IEC 62305, 2012).

Lightning causes more fatalities in India than any other natural disaster like by flood, landslide, heat stroke and cold waves. On an average, every year, 2000 people in the country die from lightning strikes (H.T., 2016). Lightning does not often mentioned in accounts of extreme weather events in India (De *et al.*, 2005). Despite its proneness to lightning fatalities, India does not have a lightning detection network. Few victims of lightning strikes actually get any official help because the phenomenon is not covered by the National Disaster Relief Fund (NDRF). Apart from that, a super cyclone hit the coastal districts of Odisha in 1999, which had taken a toll of 9893 human lives (Kalsi S. R., 2006), statistics showed that of all the calamities that befall the state at regular intervals, lightning claims more human lives than any other natural disaster. On average, 327 persons die of lightning strikes in Odisha every year (BS. 2016), accordingly. Odisha has for long been asking the union government to place lightning in approved list of natural disasters so that it could compensate for the loss of lives from State Disaster Response Fund (SDRF).

From the above discussion, it seems nationwide; less serious efforts have been taken to study thoroughly the social and administrative aspects of lightning incidents, vulnerability of areas to lightning and its mitigation. Therefore, this research study will present the compilation of incidents of lightning fatality in India as well as in Odisha so as to assess the gravity of the risks involved. Existing policy matters related to lighting hazard *vis-à-vis* natural calamities will be discussed. Mitigating measures adopted overseas will be presented and efforts necessary for the region will be suggested.

2. Methodology

Reporting of lightning deaths and injuries are haphazard and often governments have not made it mandatory. Lightning related incidents are under reported as the news about lightning deaths are gathered from newspapers, or Panchayat or medical sources or sometimes the police station (Cooper and Kadir 2010). The methodology followed in this study is based on the data collected from field survey and various sources like Bureau of Indian Standards (BIS), National Institute of Disaster Management (NIDM), National Crime Records Bureau (NCRB), district administration, Panchayat level, hospital, police station, disaster mitigation authority, relief commissioner office, newspaper, websites and journals. The corresponding results have been presented with respect to month, district, age group, and instant location *etc.*

2.1. Indian Condition

The average number of thunder and lightning days and relative hazard priorities has been presented in **Figure 2.1** (Illiyas *et al.*, 2014). As per the study and the BIS standards none of the Indian states are free from the effects of thunderstorms and lightning. According to report of the National Crime Record Bureau (RR-NCRB), 2550 people died of lightning strikes in India in 2011, 2263 in 2012, 2833 in 2013, 2582 in 2014 and 2641 in 2015. The incidence rate is 0.2 per lakhs of population. On an average the powerful bolt from the sky accounts for 2,500-3,000 lives across the country every single year. Illiyas *et al.*, 2014 compiled the data collected from online archives and fatality figures of natural disasters of NCRB reports for the period of 1967 to 2012. They have analyzed the 45 years data for 5 major calamities *viz.* flood, landslide, cold stroke, heat stroke and lightning. These calamities together have caused 1,95,745 deaths, out of which 39 per cent were due to lightning followed by flood (18 per cent), landslide (15 per cent), heat stroke (15 per cent) and cold stroke (13 per cent). The historical data gives clinching evidence that lightning is the most damaging natural disaster in India *vis-à-vis* loss of lives. Economic damage due to lightning is enormous as lightning causes damages worth billions of rupees in the housing, agriculture, industrial and public sectors. But at the administrative level this issue has not been received due importance, thereby making it an underestimated disaster or persona non grata amongst disasters. The affected families of lightning strikes are unable to receive any financial assistance from the national calamity relief pool because the thunder and lightning are not considered as natural calamities under the National Disaster Relief Fund (scroll.in, 2016).

2.2. The State of Odisha

Odisha is a state on the eastern seaboard of India, located between 17°49′ and 22°36′ North latitudes and between 81°36′ and 87°18′ East longitudes. It spreads over an area of 1,55,707 sq km. It has a 480 km coastline. Its population was 4,19,47,358 as per the 2011 census. Administratively, the state is divided into 30 districts,

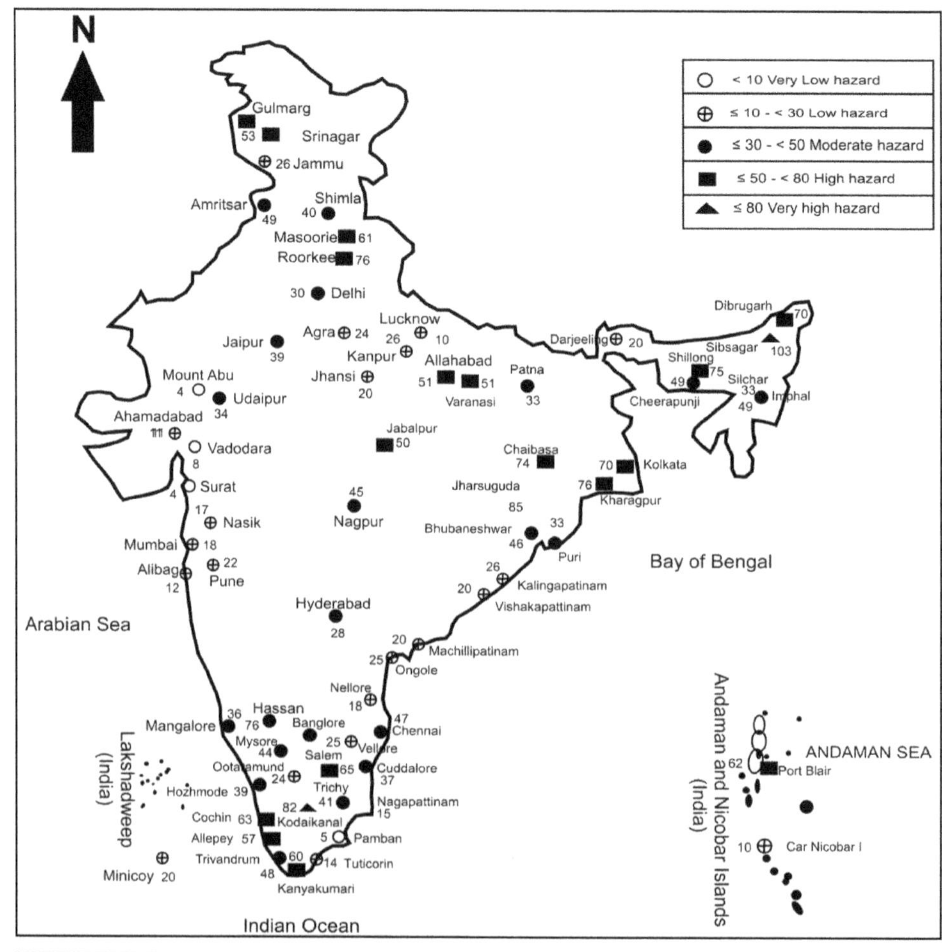

Figure 2.1: Average Number of Thunder and Lightning Days and Relative Hazard Priorities (Illiyas *et al.*, 2014).

58 sub-divisions, 314 blocks and 103 urban local bodies. The average density of population comes to 269 per sq km. with significantly higher density in the coastal areas compared to the interior parts (RR-NIDM, 2012). The State of Odisha is often called India's natural disaster capital because of the frequency with which calamities such as cyclones, floods and droughts hit the state. Apart from that, a super cyclone hit the coastal districts of Odisha in 1999, which had taken a toll of 9893 human lives (Kalsi, S.R., 2006), statistics show that of all the calamities that befall the state at regular intervals, lightning claims more human lives than any other natural disaster (Vijayalaxmi, T.N. 2013). On average, 327 persons die of lightning strikes in Odisha every year, a huge 20 per cent of the total deaths (BS 2016, TOI 2017). As many as 5235 people have been killed by lightning in the eastern Indian state in the past 16 years. Data collected from field survey for last 10 years (2007-2016) have been compiled and presented district wise in **Table 2.1**. Almost all the districts

Table 2.1: District-wise Number of Deaths Reported Due to Lightning in Odisha

Sl.No.	*Name of District*	*2007–08*	*2008–09*	*2009–10*	*2010–11*	*2011–12*	*2012–13*	*2013–14*	*2014–15*	*2015–16*	*2016–17*
1	Angul	14	18	14	12	21	18	24	14	22	19
2	Balasore	06	13	05	13	14	09	12	08	10	11
3	Bargarah	08	07	03	11	17	08	13	10	19	16
4	Bhadrakh	14	06	16	07	08	15	16	11	17	16
5	Bolangir	15	06	02	08	14	09	14	12	15	12
6	Boudh	01	05	03	02	02	02	03	02	04	06
7	Cuttack	37	14	05	17	13	15	21	17	14	16
8	Deogarah	01	02	03	03	15	08	16	11	14	13
9	Dhenkanal	38	24	09	16	34	25	32	27	31	28
10	Gajapati	02	02	01	01	00	01	02	02	04	03
11	Ganjam	22	38	12	18	28	19	23	20	27	33
12	Jagatsingpur	04	12	06	05	09	06	07	07	11	08
13	Jajpur	09	07	13	15	17	13	16	12	17	14
14	Jharsuguda	06	06	05	06	10	04	08	09	09	07
15	Kalahandi	09	05	06	07	09	05	09	07	08	06
16	Kndhamal	06	03	02	06	01	00	04	03	05	04
17	Kenderapara	18	15	15	16	10	14	18	13	16	17
18	Keonjhar	17	08	13	19	25	16	19	18	23	22
19	Khurda	07	05	10	07	21	07	12	11	22	17
20	Koraput	13	10	04	06	06	11	11	09	07	08
21	Malkangiri	10	01	02	05	04	04	06	05	08	06
22	Mayurbhanj	28	40	28	21	21	32	24	19	26	36
23	Nawarangpur	11	10	03	08	08	09	06	05	09	10
24	Nayagarah	09	05	05	05	03	04	06	04	05	07
25	Nuapada	05	06	03	04	08	01	04	05	06	05
26	Puri	05	04	02	03	07	02	03	04	06	07
27	Rayagarah	08	05	06	11	08	03	06	07	10	08
28	Sambalpur	09	14	06	07	14	09	10	08	15	16
29	Subarnpur	04	07	07	06	06	06	06	07	08	09
30	Sundergarah	15	31	14	14	22	22	21	19	21	16
	Total	**351**	**319**	**223**	**279**	**369**	**290**	**372**	**306**	**399**	**396**

have reported fatalities due to lightning. According to the number of deaths, the districts have been ranked and are shown in **Figure 2.2**. Mayurbhanj, Dhenkanal and Ganjam districts have more than 20 fatalities per year and have been top the rank. On analyzing the collected data of 13 years from various sources, it was found that mostly, the victims were male, 15 to 45 years of age, farmers, daily wage laborers

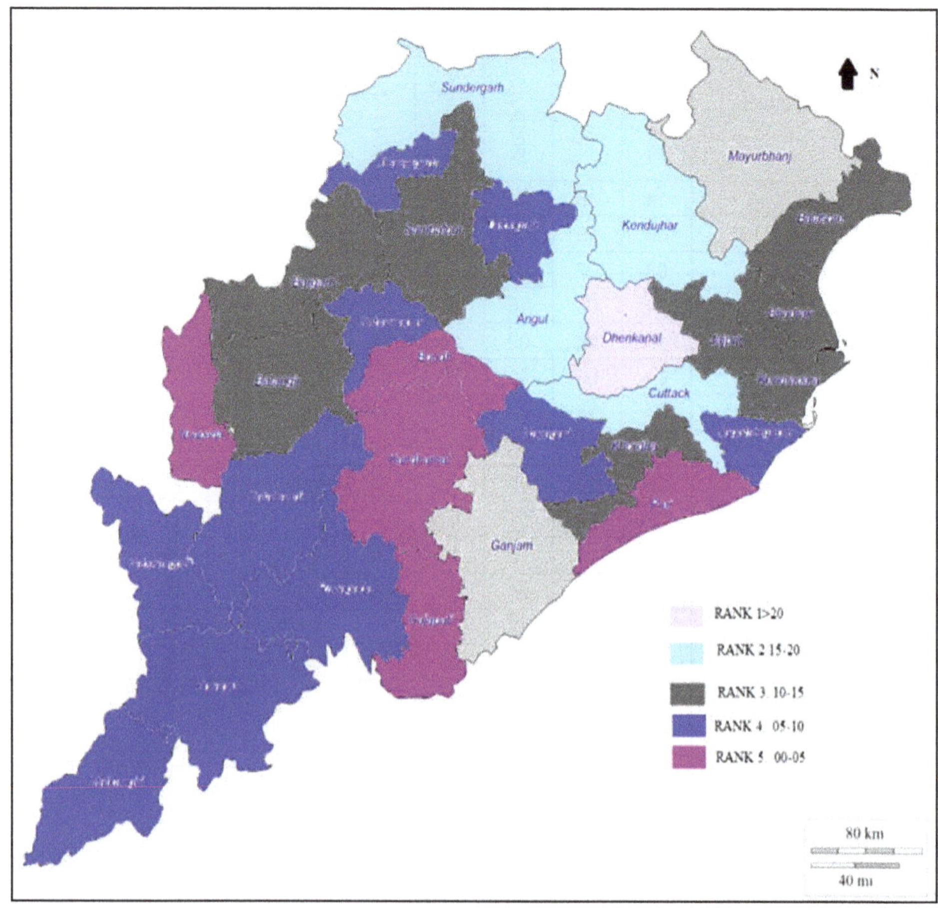

Figure 2.2: District-wise Lightening Fatality Ranks.

and people who were in the fields and were from rural areas. On an average, 74.67 per cent of those killed were male while 25.33 per cent female. Mostly, in a year, deaths due to lightning occurred during April to September with maximum of the 33.48 per cent deaths occurred in the month of June. It was followed by 31.86 per cent deaths in July. In the age group analysis, 9.26 per cent were up to the age of 14 and 57.12 per cent of those killed were between 15 to 45 years. 24.37 per cent died belong to age group between 46 and 60, while 9.25 per cent killed were aged 61 and above. At least 35.32 per cent were struck by lightning when they were working in paddy fields. Another 20.72 per cent and 13.48 per cent were struck when they were in open fields and near their house, respectively. However, 8.21 per cent people died while they were inside the house.

3. Existing Policy

As discussed earlier, few victims of lightning strikes actually didn't get any official help because the phenomenon does not come under the guidelines of

National Calamity Relief Rules (NCRR). Odisha has for long been asking the Union government to place lightning in approved list of natural disasters so that it could compensate for the loss of lives from State Disaster Response Fund (SDRF). Most of the lightning affected states in India provide an *ex-gratia* amount of one lakh Indian rupees to the kith and kin of the deceased person from the Chief Minister's Relief Fund (CMRF). CMRF has state specific guidelines to help families suffering from chronic ailments (like cancer, cardiac issues, kidney transplant, brain tumor, *etc.*), any natural or man-made calamities or accidents and a part of it is mobilized through public contribution. However, last year, the 14th Financial Commission allowed state governments to earmark 10 per cent of the State Disaster Relief Funds to provide, immediate relief to victims of disasters specific to their states, if these are not in the home ministry's notified list of disasters. Meanwhile, lightning has been declared as a 'State Disaster' by government of Odisha. As per the norms, 4 lakh Indian rupees per deceased will be paid by the government as *ex-gratia* (TOI, 2017).

4. Mitigating Measures

Despite the sense that lightning deaths are inevitable, nations such as the US have actually reduced casualties, from about 100 fatalities due to lightning in the 1970s, to the 27 in 2015 (Scroll 2016). While in the US, lightning strikes people while camping or at the beach, but in India it's mostly farmers who are most vulnerable. While the risk from lightning cannot be completely eradicated, fatalities could definitely be reduced. These can be achieved in following ways, *viz.*

(i) Local Planning and Regulatory Frameworks,

(ii) Infrastructure Facilities for Warning Systems,

(iii) Natural System Protection,

(iv) Education and Awareness Programs,

(v) Continuous Research and Development Efforts.

(vi) Separate department should be created under Disaster Mitigation Authority for policy formulation, implementation and to take responsibility and leadership activities to deal with the situation from state level to Panchayat level through districts. Elaborate guidelines shall be prepared and issued to Government and non-government organizations. Lightning risk and mitigation studies shall be included in the school and college curriculum. Capacity building measures through preparedness, warning systems, training and awareness of functionaries of all the line departments shall be carried out at State, District, Blocks and village level. Review and monitoring of different programs shall be conducted regularly during lightning intense months.

(vii) Early warning system shall be employed by putting Light Detection Networks in the field throughout the state including the weather application software for detection of lightning and storms. Based on the observations through lightning monitoring network, on time warning is issued through Hooter, SMS, Email, TV, FM Radio, mass media and also with the AgroMet advisories. People are directed to safe grids/shelters accordingly.

(viii) Lightning safety aspects need to be religiously implemented by the line Departments. Building bye-laws shall be modified to install lightning arrestors for all the structures having G+2 and more stories. Schools, Industries, Government buildings, Shelters, Hospitals, Police Stations, Paramilitary and Army Bases, Heritage Sites, Power Systems and Communication Networks, building producing and handling energetic materials, Nuclear Stockpiles and Combustible Dusts shall be made mandatory to install lightning conductors/arresters through lightning safe grids.

(ix) Creating awareness among masses for risk reduction is needed amongst policy makers, decision makers, administrators, professionals (architects, engineers and others at various levels) financial institutions (banks, insurance, house financing institutions) and NGOs and voluntary organizations. Extensive Public Awareness Programmes about safety measures shall be conducted, specially for rural masses and school children by the use of electronic and print media, school education, street play, wall paintings and by the distribution of hand outs on Do's and Don'ts of lightning.

(x) The exact correlation between Lightning incidences, geology and geomorphology of a region needs to be studied in detail with the help of sophisticated scientific equipment over a period of time. Research studies shall be carried out in to the acute climate activity and intensity of lightning, formation of cumulonimbus cloud, thunderstorms and lighting, frequency of lightning and fatalities, accident by ground conduction, population and vegetation density, topography relationships, lightning zone mapping, fatalities and economic losses *etc.* Based on frequency, intensity and severity of lightning incidences safety measures shall be suggested for adoption as per priorities.

5. Conclusion

The conclusion is that, the government should be more serious about lightning fatalities. Early warning system and mitigating efforts shall be deployed in vulnerable areas along with educating the mass through awareness program. It is time the Centre and State governments collect proper statistics and conduct research on how to reduce fatalities.

References

1. BIS 2007. Indian standard protection of buildings and allied structures against lightning- Code of Practice (second revision). National Standards Body of India. 2007

2. Business Standard 2016. http://www.business-standard.com/article/news-ians/2-297-killed-by-lightning-in-odisha-in-7-years-116090900409_1.html

3. BS EN/IEC 62305, 2012. British standard for lightning protection. Part-2 Nottingham, UK

4. Cooper, M. A., and M. Z. A. Kadir. 2010. Lightning injury continues to be a public health threat internationally. 21st International Lightning Detection Conference, 19-20 April, Orlando, USA

5. De U. S., Dube R. K. and Prakasa Rao G. S. 2005. Extreme weather events over India in the last 100 years. *J. Ind. Geophys. Union* 9: 173–187.

6. Guthrie, M. and Rousseau, A., 2014. Lightning warning techniques for risk mitigation", 10th Global Congress on Process Safety, March 30-April 02, Orlando, USA

7. Hindustan Times 2016. http://www.hindustantimes.com/india-news/why-east-india-is-prone-to-lightning/story-MV48lglt9023tsvwKu9gFP.html

8. IEEE 2010. How to protect your house and its contents from lightning. IEEE Guide for Surge Protection of Equipment Connected to AC Power and Communication Circuits. Standards Information Network, IEEE Press, USA

9. Illiyas, T.F., Mohan, K., Mani, K.S. and Pradeep Kumar P.A. 2010. Lightning risk in India challenges in disaster compensation. *Economic and Political Weekly*, Vol XlIX, 23: 23-27.

10. Kalsi, S.R. 2006. Orissa Super Cyclone – A Synopsis. *Mausam* 57: 1-20

11. Rafferty, J. P. (eds.) 2016. Dynamic Earth: Storms, Violent Winds and Earth's Atmosphere. *Britannica Educational Publishing*, pp. 249

12. NIDM 2012. Research Report- NIDM Odisha, National Disaster Risk Reduction Portal. https://nidm.gov.in/

13. NCRB 2011. Research Report- Accidental deaths and suicides in India. National Crime Records Bureau (NCRB). Ministry of Home Affairs, Government of India, New Delhi, India

14. Scroll 2016. Lightning kills more Indians than any other natural disaster. Can those deaths be prevented? Published on June 27, 2016 https://scroll.in/pulse/810598/lightning-kills-more-indians-than-any-other-natural-disaster-can-those-deaths-be-prevented

15. TOI 2017. Odisha: Lightning kills 34 people in the past 36 hours. Published on August 1st, 2017. Read more at: https://timesofindia.indiatimes.com/india/odisha-lightning-kills-34-people-in-the-past-36-hours/articleshow/59864429.cms

16. Vijayalaxmi, T.N. 2013. Lightning claims more lives in Odisha than any other calamity. *Down to Earth* Web Edition, Published on September 17, 2013.

Section II

Flood, Drought and Water Pollution

Chapter 3

The Effect of Floods on Livestock in Ampara District in Sri Lanka

E. Pavithira[1], M.G.M. Thariq[1],*
M.L.F. Ameer[2] and K. Nijamir[2]

[1]Department of Biosystems Technology, [2]Department of Geography,
South Eastern University of Sri Lanka, Sri Lanka
**E-mail: pavithira@seu.ac.lk*

ABSTRACT

The monsoonal rainfall is the predominant aspect to cause flood disaster in Sri Lanka. With respect to spatial distribution, floods are most frequently occurring in Eastern Province. Ampara is one of the three districts of the Eastern Province affected by floods with increasing losses to life and socio-economy of the inhabitants in the past few decades. Therefore, this study aims to study the economic losses on livestock. The objectives of this study were to study the mitigation measures implemented by the government to protect livestock at Ampara district and to suggest technological measures to minimize the losses by the flood disaster. Secondary data were collected from Ampara district and all the collected data were analysed. It has been found that the flood disaster in Ampara district has caused 106 million SLR (Sri Lankan Rupees) direct loss to livestock and death of animals were 21,000, 6,750, and 125,000 in cattle, goats and poultry, respectively in year 2011. Further, the indirect losses were on feed suppliers, human food security, food prices and milk producers etc. Flood disaster has caused some sort of diseases for the livestock and caused dilapidated condition of their shelters. In order to mitigate the flood effects to the livestock in Ampara district, many remedial measures have been suggested with the aid of technology.

Keywords: *Diseases, Economy, Floods, Livestock, Mitigation, Technologies.*

1. Introduction

Across Sri Lanka, climate change related weather aberrations and resultant extreme weather events or natural disasters are becoming increasingly common

(UNDP 2016). While this affects the country at large, people, farmers and agricultural workers face the worst impacts of this variability. In the past 34 years, 28 million people were affected by natural disasters. Flood is amongst the most devastating natural disaster affecting lives than any other disaster. According to the Disaster event and impact profile of Sri Lanka in 2011, 48 per cent of the people affected by floods. Major floods are associated with the two monsoon seasons. During South-West monsoon (from May to September) the Western, Southern and Sabaragamuwa provinces are vulnerable of floods. During North-East monsoon (from Dec. to Feb.) the Eastern, Northern and North Central provinces are at risk of flooding. According to the records, severe floods have occurred in Sri Lanka in the years 1913, 1940, 1957, 1967, 1978, 1989, 1992, 2003, 2007 and 2011 (Disaster Management Centre 2011).

Overall, the flooding in the Eastern Province was the worst since 1913. The most devastating floods that had battered the Eastern coast of Sri Lanka and wreaked havoc in most parts of the Island had cost the emerging economy a staggering SLR 30 billion (Sri Lanka Investment Forum, 2011). Eastern province includes three districts *i.e.* Batticaloa, Ampara and Trincomalee. Ampara District is situated in the coastal region of Sri Lanka. The conditions of climate are varied on the basis of their physical aspects. Rainfall is the main source of water available in the district of Ampara, with an average of 1600-2500 mm rainfall in a year (District Planning Secretariat, Ampara, 2010). With respect to spatial distribution, floods are most periodically occurring in the district of Ampara (UNDP 2011). This happens when the river or stream holds more water upstream than usual, and it flows downstream to the neighbouring low-lying areas through coastal region, typically referred to as the floodplains. As a consequence, this creates a sudden discharge of water into the adjacent lands of Ampara district which leading to flooding. In addition, the drainage, topography and land use patterns of Ampara district are also very prone to flood. Due to torrential rain resulting from recent weather condition, nearly 10 divisional secretariat divisions of Ampara district have been experiencing flooding since the 6th January 2008. The most affected divisions are Akkaraipattu, Aalayadivembu, Ninthavur, Thirukkovil, Pottuvil, Sammanthurai, Saithamaruthu, Karaitivu, Addalaichenai and Kalmunai (Bandara 2011).

Floods are more of common occurrence in Ampara district, heavily impacting environment, socio economy of the inhabitants, mostly the people living in the flood prone areas. Much of the assets of nation and individual have been destroyed due to the flood disaster. The increased frequency of flood incidence in the last ten years has caused severe hardship to poor farmers. According to the report stated in Flood Impact Assessment, by 19th January 2011, plain areas of Ampara affected by flash flood and highly destructive, riverine flooding has been a very destructive phenomenon, although with a slower onset, affecting densely populated cultivated areas and livelihoods. The main livelihood opportunity of the people in the district is agriculture including livestock farming (UNDP 2016).

According to mentioned above, the livestock sector plays a multiple role in the livelihood development of the people and comprises mainly with dairy, poultry, swine, goat and sheep. Ampara district through the livestock sector contribute to district GDP about 14 per cent and this district having three communities together

(Ifham 2016). There are number of families engaged in livestock in Ampara district. The livestock population of Ampara district were about 103,508 cattle, 30,687 buffaloes, 320 swine, 24,526 goats and the annual milk production was 22,033,244 litres (DAPH, 2015). Poultry sector is a very important livestock sector in Ampara district. 20,941 families in Ampara district are involved poultry rearing.

The need to study the effects of floods on livestock has begun since flooding has become a problem to society, people and their valuables. Historically many solutions have been proposed to mitigate the effects of flooding by various governments across the world. Mitigating flood effects requires information on the flooding characteristics and how such characteristics propagate. Therefore, the present study evaluated and assessed in detail to achieve primary objective of studying the economic losses on livestock. Consequently as secondary objectives to study the mitigation measures implemented by government which are also helpful to protect livestock at Ampara district and to suggest technological measures to minimize the losses by the flood.

2. Methodology

This study is about the effect of floods on livestock in Ampara district as study area. It is located in eastern province of Sri Lanka and situated within the range of the north latitude 7° 25′ to 7° 27′ and the east longitude 81° 45′ to 81° 50′ experiencing flood events frequently during North-East monsoonal season. The MSL (*mean sea level*) of this area is 9 m. Annual rainfall 1577 mm registered in November and December due to the North-east monsoon. The major economical activities in this region were farming and animal husbandry. The study area in Ampara district consists of Thirukkovil, Sainthamaruthu, Pothuvil, Ninthavur, Karativu, Lahugala, Kalmunai, Alayadiwembu, Akkaraipattu and Addalachchenai divisional secretariat Divisions.

Secondary data were collected from reports of Ampara District secretariat, Disaster Management Centre and Department of Irrigation (Ampara regional office), magazines about Sri Lanka's Floods, previous researches, websites from various sources. The result was derived using the analysis of various data obtained from above sources.

3. Results and Discussions

3.1 Economic Effects on Livestock due to Floods in 2011

In Sri Lanka, most specifically the unparalleled monsoonal rains that continued flooding in the eastern region in the months of January and February andcaused its harmful impacts on livelihood. The most affected districts were Batticaloa and Ampara. **Figure 3.1** shows the estimated livestock damages due to floods in January to February period in 2011.

Results appeared in the **Figure 3.1** shows that the damage to livestock has been estimated as SLR 1,230 million (nearly 92 per cent) due to deaths of cattle, buffalos, goats, poultry, other livestock and infrastructure damagesin Batticaloa district. In Ampara district, damage to livestock is estimated as SLR 106 million (nearly 8 per

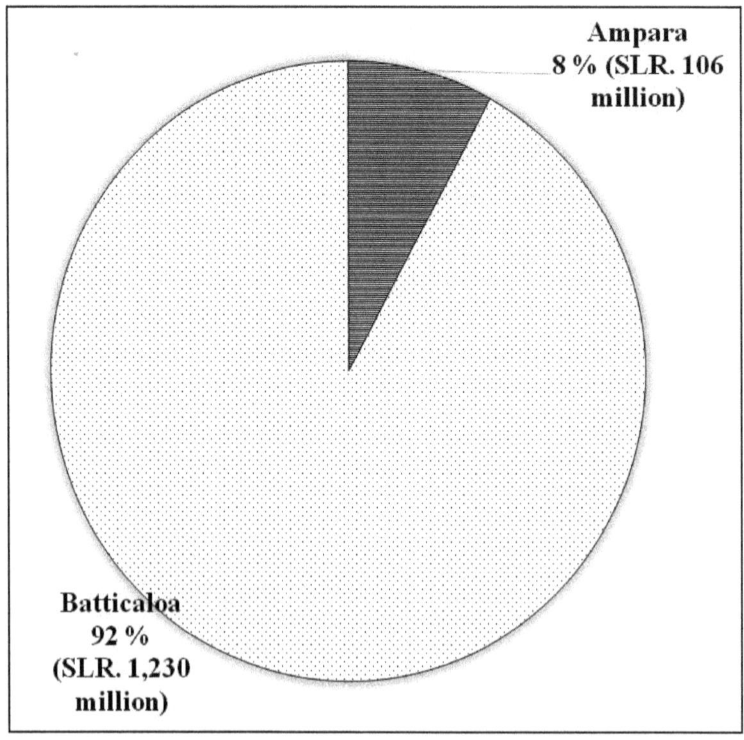

Figure 3.1: Economic Effects on Livestock Due to Floods in 2011.

cent). These are initial estimates handed over by the Provincial Line Ministries and Directors of Planning in the affected districts.

3.2 Livestock Damages due to Floods in 2011

The potential impacts on livestock directly and indirectly include changes in production, quality of feed crop and forage, water availability, animal growth, milk production, diseases and biodiversity *etc.* These impacts are primarily due to the effects of floods.

Figure 3.2 shows the damages to livestock due to floods during 26th December 2010 to 14th January 2011. The number of animals were severely affected in eastern province. Ampara is the second most affected district brought higher economic damages to livestock next to Batticaloa. About 21,000 cattle, 6,750 goats and 125,000 poultry were reported dead. Livestock was severely impacted with an average of 40 per cent of livestock lost by flood affected households. As a consequence, it increased the cases of diseases such as dengue, pneumonia and other flood-induced sicknesses. It reduced quantity of food available due to livestock death and injuries during disaster. Therefore, Department of Animal Production and Health has requested emergency veterinary medicine to effectively treat the large numbers of livestock suffering from the floods (United Nations 2011).

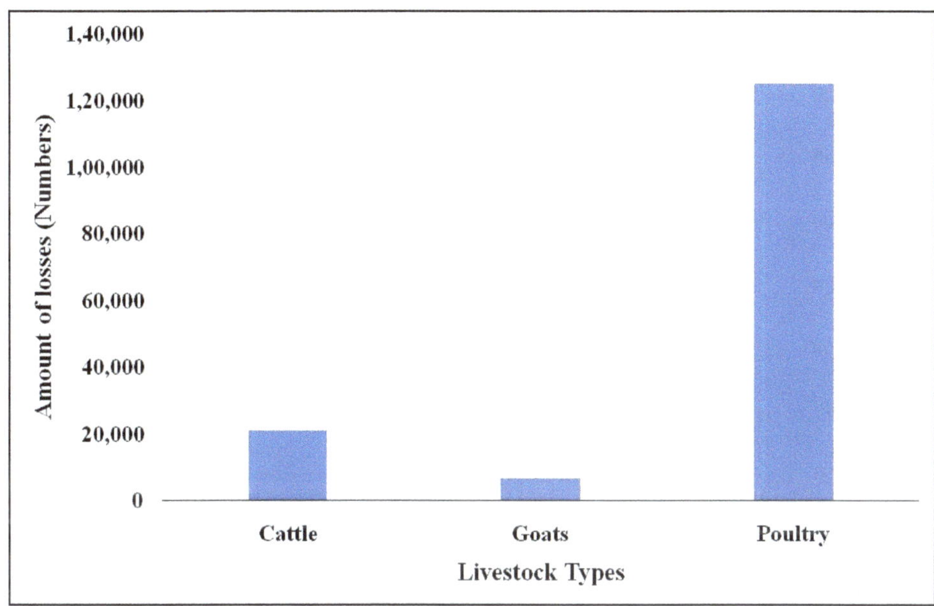

Figure 3.2: Livestock Damages Due to 2011 Flood.

Meanwhile, the impact of floods on livestock have the potential to indirectly affect feed suppliers and markets, reduce human food security, reduce ability of governments and firms to provide services to the public, rising food prices, reduce wages of milk producers, loss of livestock as an input to some industries, reduce availability of manure for agricultural farmers and increase level of malnutrition due to reduced animal sourced foods *etc.*

3.3 Livestock Damages Due to Floods

The data appeared in **Figure 3.3** shows, the damage to overall livestock sector is not significant, although poor families have lost their domestic livestock. The largest damages were reported from Ampara district compared to Colombo, Kalutara, Galle, Batticaloa, Trincomalee, Mulaitivu, Kilinochchi and Jaffna. It is about 29,520 poultry, 3,420 cattle and 1,110 goats were reported dead. Pothuvil milk collecting centre and veterinary dispensary were fully destroyed. This is significantly higher than the estimated damage to the crop fields (World Bank 2005).

3.4 Identified causes for Flooding in Ampara District

Floods are natural occurrences where an area or land that is normally dry abruptly becomes submerged in water. Many conditions result in flooding in various regions across the globe. Here are some leading causes of flooding in Ampara District.

1. ***High Rainfall Intensity:*** The most common cause of flooding is due to intensive rainfall or rainfall for an unusually long period of time.

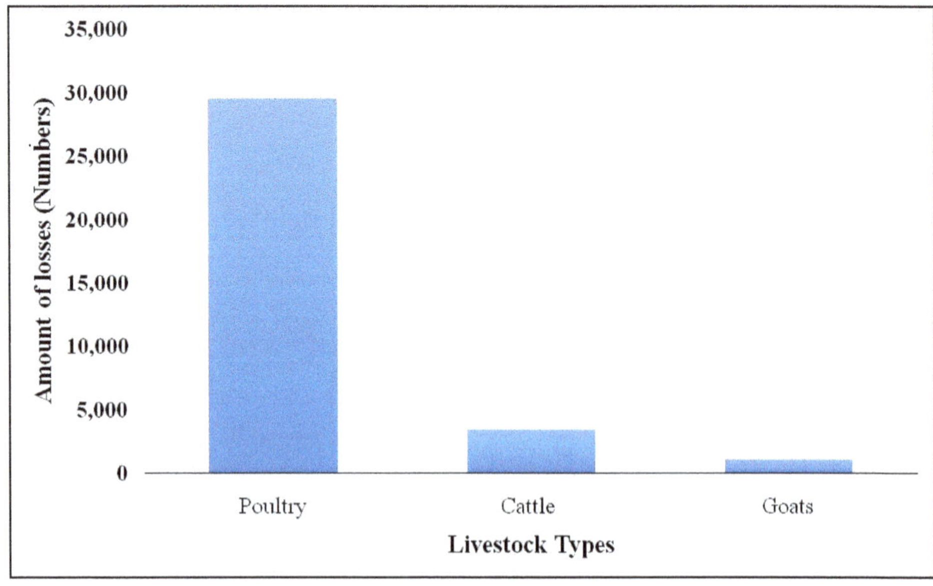

Figure 3.3: Livestock Damages Due to Floods in 2004.

Ampara is such an area receiving high rainfall intensity due to North East monsoonal season. Thereby intensifying the risk to flood occurrence.

2. ***Poor Drainage Management:*** There was an inadequate drainage to carry away surface water quickly.
3. ***Lack of Maintenance of the Canals:*** Poor maintenance of drains and irrigation canals has reduced the flow carrying capacity of the drainage system. Thus triggered flash floods in Ampara.
4. ***Deterioration of Wetlands:*** Insufficient awareness among people about the wetlands caused the wetland damages resulting in inability to store flood water and maintain surface water properly.
5. ***Increased Runoff:*** If it rains for a long time, the ground will become saturated and the soil will no longer be able to store water leading to increased surface runoff.
6. ***Rapid Urbanization:*** Increased population and urbanization brought wetland deterioration and extinction due to lack of awareness among people. It leads to the changes in water level.
7. ***Improper Town Planning:*** Proper town planning will ensure that waterways are not blocked. This way, water can quickly run through if it rains and minimize any chance of town flooding. However, lack of town planning in Ampara District because flooding.
8. ***Dam Breakage:*** Dams are man-made structures used to hold water from flowing down from a raised ground. The potential energy stored in the dam water is used to generate electricity. At times, the walls can become

weak and break because of overwhelming carriage capacity. Due to this reason, breakage of the dams caused extensive flooding in the adjacent areas in the Ampara district.

3.5 Mitigation Measures Implemented by the Government

3.5.1 Drainage Improvement Projects

Poor drainage is one of the identified causing agents for flood occurrence. Drainage improvement projects were implemented by the government in Ampara district to improve the drainage which is estimated to cost 117.6 million SLR. Surface water were carried away quickly due to the adequate drainage arrangements after the drainage improvements projects. Implemented projects are mentioned in **Table 3.1.**

Table 3.1. Drainage improvement projects in Ampara District (DMC 2011)

Akkaraipattu	Rehabilitation of 4 km length of Tillar Uppalam drainage, Mura Odairetention ponds and Kudimutturei drainage.
Alayadivembu	Drainage canal improvement.
Uhana	Demolition of sand bunds and drainage canal improvement.
Pothuvil	Excavation and rehabilitation of silted sinhapura Thona.

Modification of the drainage channel are performed to improve the conveyance lead to increase flow velocities, which in return results in reduced flood stages or water surface elevation for a given flood at a location. The result is in agreement with Ahn (2015) who simulated alternative solutions to structural measures used during flood drainage and analyzed the efficacy of flood mitigation measures.

3.5.2 Flood Mitigation Projects

By analysing of past damages, response, relief and recovery data of previous disasters in Ampara and Batticaloa districts, DMC has identified the need for a flood mitigation plan. In order to reduce flood risk in those districts DMC developed a comprehensive flood mitigation programme called a "*Comprehensive Study on Flood Mitigation in Ampara and Baticaloa Districts*" with the financial assistance of UNDP. The study was completed in April 2010 and the findings and recommendations were discussed with respective officials in both the districts as short term and long term measures to minimise effect of flood. Fund allocated for flood mitigation projects in Ampara and Batticaloa as SLR. Mn 25. But the weakness is, it is still in progress.

3.5.3 Disaster Management Plans

In each and every district, District Disaster Management Coordination Committee (DMC) is the main management body. The members of the committee, the sub committees and their roles and responsibilities are listed in the plan. Also these plans contain base line data of the administrative area, risk and vulnerability assessment for the hazards prevailing in the area, contact details of focal points, resources available in the area and responsibilities of different stakeholders involved in disaster response *etc.* These plans have been distributed among the members of District Disaster Management Coordination Committee for their usage during a disaster. Therefore, the document plan which gives the district mechanism for responding to disasters if and when a disaster occurs. According to the DMC annual report, 2010, the draft document final plans for Moneragala, Trincomalee, Batticaloa and Ampara Districts are ready with final proof reading.

3.5.4 Awareness among Public for Upcoming Monsoon by DMC

Two workshops were organized focusing on the required preparedness and responsibilities of respective agencies to face the monsoon season. The first was held on the 24th September 2010 in Batticaloa with participation of 47 members from different agencies of Ampara, Batticaloa and Trincomalee districts (DMC Annual Report, 2010). Most of the programmes were organized and conducted by DMC in collaboration with UNDP and other government and non-governmental stakeholders. It facilitated the awareness among public for upcoming monsoon.

3.5.5 Livestock Insurance Schemes

Government introduced the *Livestock Insurance Schemes* (LIS) to compensate for loss of animals or reduced productivity due to a natural hazard or climate disaster. It enhances the need to raise awareness of existing insurance programmes and their benefits among farming communities. Total sum of LIS payment for Ampara District due to the flood disaster was about 37.1 million SLR in 2011 flood (**Table 3.2**).

Table 3.2: Compensation of Payment for Ampara District Due to the Flood Disaster

Estimated Compensation Payment (in million SLR)			*Total*
Buffalo	*Goat*	*Poultry*	
13.0	11.8	12.1	37.1

But participation and corporation of the people in the community is very essential for the economic development of the district. Thus, the people should think that this is our property and we have to contribute to our best for the betterment of the district and country as a whole.

3.5.6 Activities for Disasters Related to Large Dams

Dam failures are often caused by over-topping of the bund due to inadequate spillway capacity during large inflow to the reservoir from heavy precipitation. DMC has initiated a programme to improve the preparedness capacity of communities in downstream of selected major dams considering the possibility of dam during heavy rain to mitigate floods. The program consisted of awareness programs for downstream vulnerable communities forming subcommittees such as first aid, camp management and village security, development of early warning dissemination mechanism, mapping of inundation areas and displaying signboards. Dam safety programs were carried out in Ampara district also. Allocated cost for that dam safety program in Ampara as SLR 15000 (DMC Annual Report 2014).

3.5.7 Development of Flood Awareness Creation Materials

Awareness creation materials on Disaster Risk Reduction (DRR) were collected from agencies engaged with the development of DRR materials and gaps were identified. Based on the requirement materials were developed with the assistance of relevant technical agencies. This includes a cartoon booklet and an awareness booklet on floods.

4. Conclusions

Ampara district is considered as a densely populated and cultivated area in Eastern province badly affected by flood devastation. The occurrence of flooding seems to be most periodic in the latter years, with the most flooding occurred in the year 2011 in Ampara District. However, during the flood, this district suffered huge set back interms of the economic development. Livestock have been heavily affected directly and indirectly such as goat, cattle, buffalo, poultry and cattle. People have become increasingly lost their livestock by floods with the highest number recorded in 2011. However, the occurrence of economic damages due to floods were quite highwhere it recorded of 21,000 cattle, 6,750 goats and 125,000 poultry. It has been found that the flood disaster in Ampara district has caused 106 million SLR direct losses to livestock in 2011. Thus, in the light of above analysis following conclusions can be drawn from this study.

- ✰ Mitigation measures such as drainage improvement projects and district disaster management plans were implemented by government to minimize flood hazards.
- ✰ DMC and UNDP has also increased development projects on floods.
- ✰ Government introduced the livestock Insurance schemes to compensate for loss of animals or reduced productivity.
- ✰ Some measures were suggested to protect livestock due to floods in Ampara district
- ✰ Technological measures were suggested below as recommendations with aid of technology such as flood plain modelling management, hazard mapping using GIS, emergency request for satellite observation, usage of special devices to monitor rainfall data in real time *etc.*

5. Recommendations

5.1 Non-technological Measures to Minimize the Losses by the Flood

- ✰ Providing veterinary information and services to livestock farmers, especially during disaster situations.
- ✰ Establishing disaster management committee.
- ✰ Improve the health facilities and disease control of the livestock with experts.
- ✰ Build institutional and individual capacity of government officials, local extension workers and farmers on disaster preparedness for livestock sector.

- Control and maintain livestock densities for optimal output considering space requirement during floods.

5.2 Technological Measures to Minimize the Losses by the Flood

- Establish early warning systems for identifying risks to livestock sector in collaboration with Product Development and Management Association (PDMA) and research institutions.
- Promotion of biotechnology to produce improved livestock using genetic engineering to tolerate floods.
- Flood plain modelling management is used to plan communities and to test the effect of potential flood water on a community, while the mitigation side incorporates procedure for flood damage reduction and overall watershed management.
- Usage of special devices to monitor the accurate rainfall data in real time. For this, an International Water Management Institute (IWMI) scientist has invented a simple device that can monitor the accurate rainfall data in real time.
- Implementation of Emergency Request for Satellite Observation from Japan Aerospace Exploration Agency (JAXA) along with the Sentinel Asia and ADRC provides facilities to request satellite images during emergency situations.
- Hazard mapping using GIS software to guide and identify the suitable location for flood relief centre in the flood prone areas during the flood events.
- Use GIS integrated modelling applications to manage large volumes of geospatial data to provide spatially distributed parameters for modelling.
- To avoid loss of life is by issuing a flood warning to the public living downstream of the dam under thread.
- Implement a system for livestock farmers to get quick messages through mobile phones before the occurrence of heavy rains or floods.

References

1. Ahn J. 2015. Flood Mitigation Analysis for Abnormal Flood at Namhangang River Basin. Master's Thesis. Ajou University, Suwon, Korea.
2. Bandara K.A.J.M. 2011. Flood damages in Ampara district 2011: Engineering hydrology, Faculty of engineering, University of Ruhuna,Matara, Sri Lanka
3. DAPH 2015. Livestock Statistical Bulletin 2015 [Online]

   Available at: http://www.daph.gov.lk/web/images/content_image/publications/other_publications/stat_bulletin_2015.PDF (Accessed 25th October 2017).
4. Disaster event and impact profile 2011. Disaster Emergency Situation in Sri Lanka: Eastern Province and Pollonnaruwa District [Online].

Available at: http://www.dmc.gov.lk/attchments/Presentation%20-0Floods%20in%20Batti%20 and%20Amp-%20Updated%20on%2014-1-2011.pdf (Accessed 25th September 2017).

5. Disaster Management Centre 2011. Recent Floods in Sri Lanka. [Online].

   Available at: http://www.adrc.asia/acdr/2011colombo/documents/01_Recent%20Floods%20in%20Sri%20Lanka%20-%2010.06.2011.pdf (Accessed 25th September 2017).

6. Disaster Management Centre 2014. Annual Report [Online].

   Available at: http://www.dmc.gov.lk/Annual%20Plans/Annual%20Report-%202014.pdf (Accessed 29th September 2017).

7. District Planning Secretariat, Ampara. 2010. Consolidated District Annual Implementation Programme, District secretariat office, Ampara, Sri Lanka

8. Ifham Nizam 2016. Poultry sector developing in Ampara: The Sunday Leader. [Online]

   Available at: http://www.thesundayleader.lk/2016/11/20/poultry-sector-developing-in-ampara/ (Accessed 10th October 2017).

9. Sri Lanka Investment Forum 2011. Asian Tribune to organize Sri Lanka Investment forum. [Online].

   Available at: http://www.asiantribune.com/News (Accessed 20th October 2017).

10. UNDP 2016. Coping with Climate Change and Variability: Lessons from Sri Lankan Communities. [Online].

    Available at: https://reliefweb.int/sites/reliefweb.int/files/resources/Coping%20with%20Climate%20Change%20and%20Variability.pdf (Accessed 30th September 2017).

11. UNDP 2011. Sri Lanka Comprehensive Disaster Management Programme [Online].

    Available at: https://info.undp.org/docs/pdc/Documents/LKA/SS2CDMP%20Pro.%20doc%20final.docx (Accessed 30th September 2017).

12. United Nations 2011. Flash Appeal Revision [Online].

    Available at: https://www.humanitarianresponse.info/system/files/documents/files/revision_2011_sri_lank_fa_screen.pdf (Accessed 25th September 2017).

13. World Bank 2005. ANNEX VII- *Agriculture and livestock* [Online].

    Available at: http://siteresources.worldbank.org/INTSRILANKA/Resources/2330241107313542200/slna-annex07.pdf (Accessed 25th September 2017).

Chapter 4

Effect of Flood on the Out Breaking of Dengue Fever in Sri Lanka in 2017

M.B.F. Jemziya

Department of Biosystems Technology, Faculty of Technology, South Eastern University of Sri Lanka, Sri Lanka

E-mail: fathimajemziya@yahoo.com

ABSTRACT

Sri Lanka is frequently affected by flooding, which enhances the threat of dengue fever and impacts destructively on the nation's socioeconomic and developmental progress. Sri Lanka obtains most of its rainfall during the southwest and northeast monsoonal seasons. In 2017 immense rainfall during the southwest monsoon season initiated serious flooding, mainly affecting the Western Province, particularly the Kalutura district. This event was named the 2017 Sri Lanka floods. Number of dengue fever cases was reported, with 0.89 per cent of population being notified as dengue-affected people. The Western Province was the highest proportion of its population affected by dengue fever (1.51 per cent). Dengue fever outbreaks can be reduced by destroying the breeding habitats of vector mosquitoes by proper environmental management techniques, alteration and maintenance of urban infrastructure, and community awareness programs.

Keywords: *Dengue fever, Disasters, Flooding, Southwest monsoon, Sri Lanka.*

1. Introduction

Sri Lanka is an island nation in the Indian Ocean, southeast of the Indian subcontinent, in a strategic location near major Indian Ocean sea lanes. In recent years Sri Lanka has experienced several natural disasters with significant loss of human life and damages to property (NITF, 2016). Floods, landslides, cyclones, droughts, wind storms, coastal erosion, tsunami, sea surge, and sea-level rise are the major natural hazards, causing loss of life and enormous damage to property. In addition to these natural disasters, the country also incurs heavy losses on account of manmade hazards such as deforestation, indiscriminate coral, sand and gem mining, and industrial hazards, ethnic conflicts and occasional political violence (Sri Lanka Disaster Knowledge Network, 2009). However, the greatest threats to Sri Lanka are weather-related hazards because the island lies in the path of two monsoons (Ministry of Disaster Management, 2017).

Floods are the most commonly occurring disaster in Sri Lanka. Flooding may occur as an overflow of water from water bodies, such as a river, lake, or ocean due to the accumulation of rain water. The increase in population and consequent need for land has forced more and more people to live and work in vulnerable areas, thereby intensifying the risk to life and property in the event of major floods. Heavy rainfall, the large volume of runoff water from the catchment areas of rivers, deforestation, improper land use and the absence of scientific soil conservation practices are the major factors contributing to floods in Sri Lanka. Urbanization, coupled with insufficient drainage systems, triggers urban flash floods. This is compounded by global phenomena, like climate change, that increase rainfall intensities (Sri Lanka Disaster Knowledge Network, 2009).

The weather in Sri Lanka is dictated by two monsoon seasons that bring rain to the west and southwest coasts and the northern region of the east coast. The southwest monsoon typically peaks during late May to the beginning of June, with prevailing winds from the south and southwest, streaming toward the Bay of Bengal. The areas that usually receive the heaviest rain are in the south and west of the country, including *Kalutara*, *Ratnapura*, and *Colombo*. The flood resulted from a heavy southwest monsoon which was worsened by the arrival of the precursor system to Cyclone Mora (UNDP 2017) causing flooding and landslides throughout Sri Lanka in 2017 (BBC 2017). The event was named the '***Sri Lanka floods of 2017***'. The floods affected 15 districts in the Western, Sabaragamuwa and Southern Provinces and part of the Central Province (The Guardian 2017). The worst-affected districts were *Kalutara* (BBC 2017), *Matara* and *Ratnapura* (CNN 2017), where many people were displaced and their properties partially or completely damaged (Al Jazeera 2017).

Floods impact on both individuals and communities, and have social, economic, and environmental consequences. The immediate impacts of flooding include loss of human life, damage to property, destruction of crops, loss of livestock, and deterioration of health conditions (Queensland Government 2018). Weakening of health status of community is an alarming factor resulted from flooding. Flooding is associated with an increased risk of infection (water-borne diseases) and may indirectly lead to an increase in vector-borne diseases through the expansion in the number and range of vector habitats. Malaria, dengue and dengue hemorrhagic fever, yellow fever, and west Nile fever are considered common vector-borne diseases (WHO 2018a).

Dengue fever is one of the most serious diseases in world, particularly in Sri Lanka. Severe infection or late diagnoses are often fatal, particularly for children and susceptible persons. Water that has been stored in a container for a long period, extended rainfall during the last rainy season, and high ambient relative-hun[illegible]y and temperature may favor the breeding of *Aedes* mosquitoes (Gubler 1998, Gubler 2002, Guzman and Kouri 2004), which is the vector of dengue fever. Flooding is the most important factor affecting the habitat and life cycle of the *Aedes* mosquito. Communities devastated by the flooding face often face another battle with disease (Relief Web 2017). The present study evaluates and assesses in detail the emergence of dengue fever after flooding and its impact on human health. We also suggest measures to mitigate outbreaks of dengue fever.

2. Method and Data Collection

Secondary data was collected for this study from the *Disaster Management Center* (DMC), National Dengue Control Unit, Epidemiology Unit, Ministry of Health and various research papers, newspaper and magazines. The data were analyzed to determine the correlation between flooding and the occurrence of dengue invulnerable districts in Sri Lanka.

3. Outcome and Discussion

3.1. Occurrence of Dengue Fever Following the 2017 Sri Lanka Floods

Better health is central to human happiness and well-being. It also makes an important contribution to economic progress, as healthy populations live longer, are more productive, and save more (WHO, 2018). The outbreak of dengue fever following flooding is a significant challenge for Sri Lanka. The proportion of the population contracting dengue fever from 2010-2017 is shown in **Figure 4.1**. The number of people affected by dengue fever from Jan.-Dec. in 2017 is shown in **Figure 4.2**. There is a strong peak in July (41,121 cases). The peak coincides with

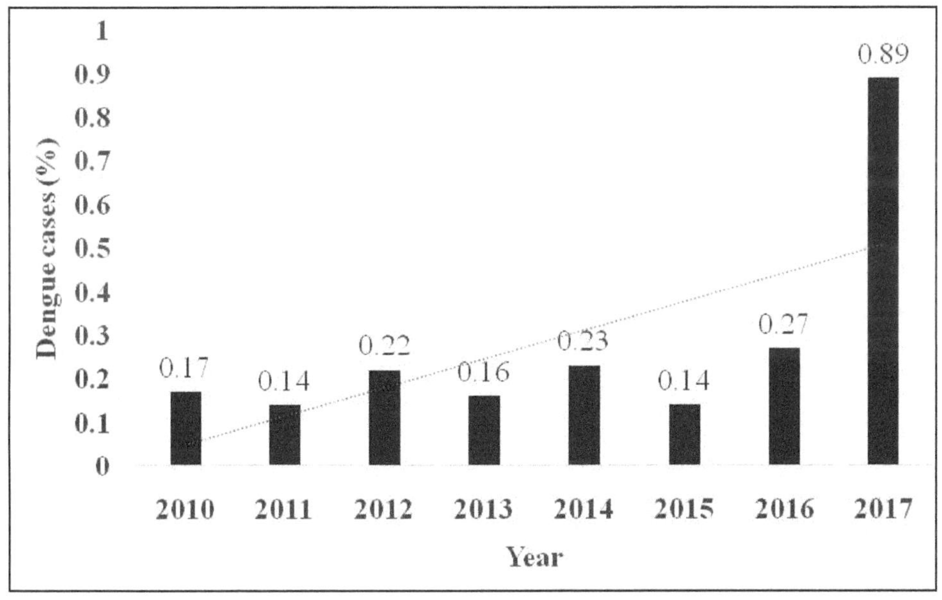

Figure 4.1: Percentage of Affected Population (Cases) by Dengue Fever in 2010-2017.

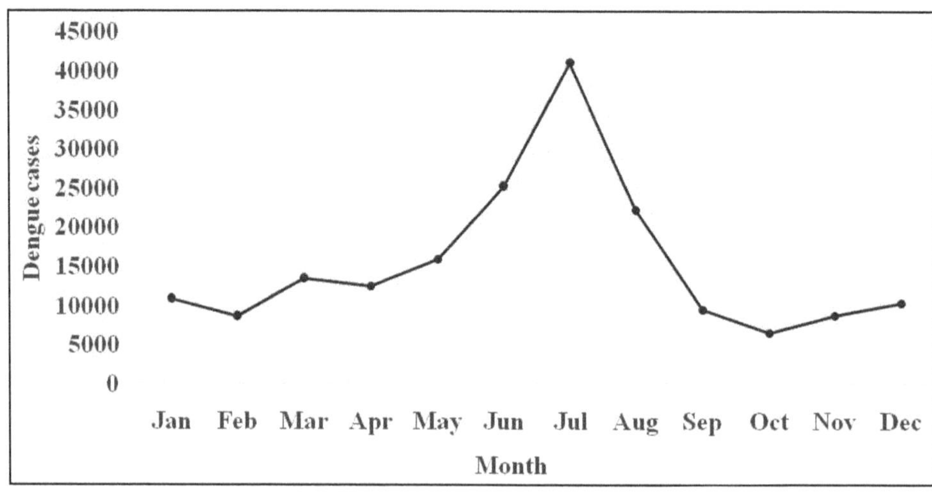

Figure 4.2: Number of People Contracting Dengue Fever Jan.-Dec. 2017.

the southwest monsoon period (May-Sept.) and the 2017 floods. The Epidemiology Unit, Ministry of Health stated that 61.48 per cent of dengue cases in 2017 occurred in the five-month period May to Sept.

Rainfall in Sri Lanka has multiple origins. Monsoonal, Convectional and Expressional rain accounts for a major share of the annual rainfall (Department of Meteorology 2016). The mean annual rainfall varies from year to year. Excess rainfall leads to stagnant water bodies and overflow. This occurrence supports to aggregate the out breaking of *Aedes* mosquito to the environment. The data illustrated in this **Figure 4.1** that flooding significantly affects the health of community. In 2017, dengue cases were notified comparatively highest (0.89 per cent) than the recent past years. According to the data of epidemiology unit, around 185,688 people were affected out of total population (20.88 million) by dengue fever in 2017. Therefore, Sri Lanka flood 2017 was a greatest resource for dengue fever in Sri Lanka. This shows that 2017 Sri Lanka flood was the major reason to evolving the outbreak of *Aedes* mosquitoes and dengue fever to Sri Lankan communities by provision of breeding habitat for vector mosquitoes and ambient environment.

3.2 Areas Vulnerable to Flooding and cases of Dengue Fever

The number of people in each province affected by dengue fever in 2016-2017 is shown in **Figure 4.3**. The dengue fever outbreak occurred in the context of massive heavy rains and flooding. Heavy monsoon rains, public failure to clear rain-soaked garbage, standing pools of water and other potential breeding grounds for mosquito larvae contribute to the higher number of cases reported in urban and suburban areas of Sri Lanka (WHO 2017). The Western Province had the most cases (88,043) in Sri Lanka. In 2017, 1.51 per cent cases that occurred in the Western Province were notified by the Epidemiology Unit, Ministry of Health, Sri Lanka. The Western Province is vulnerable to recurrent flooding as a result of an increase in

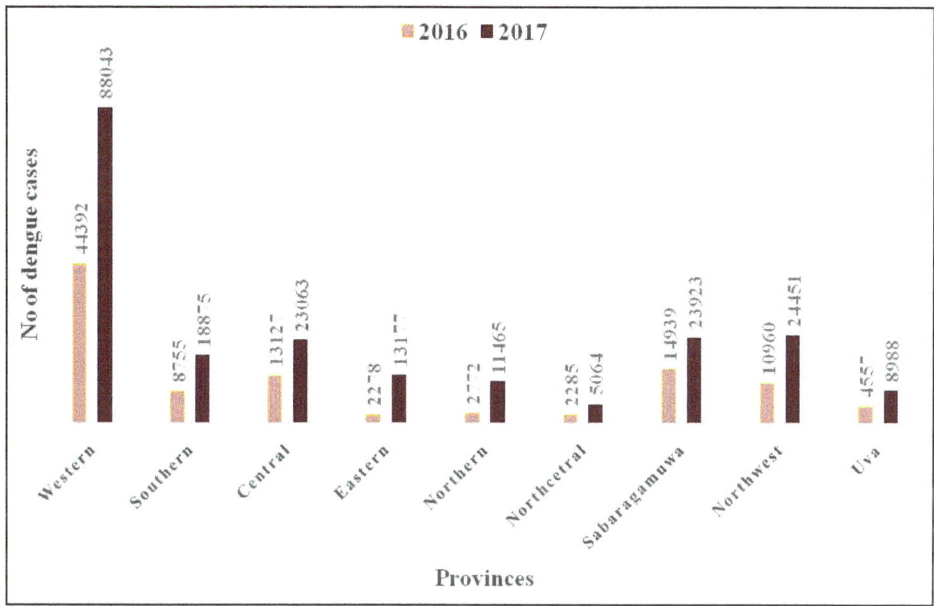

Figure 4.3: Dengue Fever Cases Prevailing in Provinces (2016-2017).

average rainfall coupled with heavier rainfall events, with knock-on impacts to the infrastructure, utility supply and the urban economy of the Province. As the most urbanized province in Sri Lanka, these climate events pose a number of problems due to the rapid urban growth the province has undergone (CDKN 2014).

The 2017 Sri Lanka flooding also severely affected the Sabaragamuwa Province (23,923 cases), Southern Province (18,875 cases) and part of the Central Province (23,063 cases).The number of noticeable dengue cases in 2017 is significantly greater than in 2016.

4. Precaution and Mitigation Measures

The government, private sectors, NGOs, and public agencies need to work together to protect the public from dengue after the rainy season. Therefore, they schedule several strategies, actions and implement to overcome and minimize the dengue hazards.

4.1. Environmental Management

Environmental management seeks to change the environment in order to prevent or minimize vector propagation and human contact with the vector-pathogen by destroying, altering, removing or recycling non-essential containers that provide egg, larval and pupal habitats. Such actions should be the mainstay of dengue vector control.

The *World Health Organization* (WHO) classifies the environmental management techniques into environmental modification, manipulation and changes to human habitation or behavior. ***Environmental modification:*** deals with long-lasting physical transformations to reduce vector larval habitats (*e.g.* piped water supply). ***Environmental manipulation***: deals with temporary changes to vector habitats involving the management of essential containers (*e.g.* water storage vessels). ***Changes to human habitation or behavior***: refer to protocols to reduce human–vector contact (*e.g.* mosquito screens).

4.2. Community-based Control Programs

Community-based control programs are developed with the aim to educate the community about the measures for the extermination of mosquito breeding sites and early detection of dengue fever. The significance of community-based programs for elimination of dengue mosquitoes has been proven in the form of elevated awareness among the communities (Rather *et al.*, 2017).

4.3. Improvements in and Maintenance of Urban Infrastructure and Basic Services

The WHO lists a number of actions that can be taken to mitigate the risk of dengue fever (WHO 2018). These are: **(i)** improvement of water supply and water-storage systems, **(ii)** mosquito-proofing of water-storage containers, **(iii)** solid waste management, **(iv)** Street cleansing and **(v)** modification or reduction of potential larval habitats while planning and constructing buildings and infrastructures.

5. Conclusions

Sri Lanka is an island of the Indian Ocean. It has seasonal variation within a year due to its geography and locality in the latitude of world. The southwest and northeast monsoons are responsible for massive rainfall, and have the greatest impact on Sri Lanka's socioeconomic and agriculture sector. Unfortunately, these rains have a negative impact on the nation's economy and health. Sri Lanka is frequently affected by flooding, most recently by the Sri Lankan floods of 2017. Following the flooding, several serious health issues emerged, chiefly an outbreak of dengue fever. Stagnant water is essential for the life cycle of the *Aedes* mosquitoes. Thus floods provide a huge impetus to dengue outbreaks in vulnerable areas. According to the statistics, noticeable dengue fever cases increased dramatically in 2017 compared to preceding years. Dengue fever peaked in the May to September, coincident with the southwest monsoon. The floods mostly affected the Western Province and this is reflected in the dengue statistics.

The WHO has played a prominent role in developing control and mitigation measures available to the government and private sectors. Environmental management techniques, alteration and maintenance of urban infrastructures, and community awareness programs can help to eliminate dengue fever in Sri Lanka by reducing the breeding habitats of the vector *Aedes* mosquito.

References

1. Al Jazeera. 2017. Sri Lanka floods leave 600,000 people displaced.

   http://www.aljazeera.com/news/2017/05/sri-lanka-floods-leave-600000-people-displaced-170531132021197.html (Retrieved 15th January, 2018)

2. BBC News. 2017. Sri Lanka floods: Scores die as monsoon triggers mudslides.

   http://www.bbc.com/news/world-asia-40063400 (Retrieved 15th January, 2018)

3. CDKN Climate and development knowledge network. 2014. Inside story: Integrating urban agriculture and forestry into climate change action plans: Lessons from Sri Lanka.

   https://cdkn.org/resource/integrating-urban-agriculture-and-forestry-into-climate-change-action-plans-lessons-from-sri-lanka/?loclang=en_gb (Retrieved 16th January, 2018)

4. CNN 2017. Sri Lanka floods: Battle to rescue stranded as death toll tops 180.

   http://edition.cnn.com/2017/05/29/asia/sri-lanka-floods/index.html (Retrieved 15th January, 2018)

5. Department of Meteorology, Sri Lanka 2016. Climate of Sri Lanka. Jafna, Sri Lanka

6. Gubler D.J. 2002. Epidemic dengue/dengue hemorrhagic fever as a public health, social and economic problem in the 21st century. *Trends in Microbiology* 10: 100–103

7. Gubler D.J. 1998. Dengue and dengue hemorrhagic fever. *Clinical Microbiology Reviews* 11: 480-496

8. Guzmán M.G. and Kourí G. 2004. Dengue diagnosis, advances and challenges. *International Journal of Infectious Diseases* 8: 69–80

9. Ministry of Disaster Management 2017. Hazard profile of Sri Lanka.

   http://www.disastermin.gov.lk/web/index.php?option=com_content and view=article and id=57%3Ahazard-profile-of-sri-lanka and catid=73%3Areports and Itemid=70 and lang=en (Retrieved on 12th January, 2018)

10. NITF 2016. National Insurance Trust Fund, Ministry of National Policies and Economic Affairs. Colombo, Sri Lanka.

11. Queensland Government 2018. What are the consequences of floods?

    http://www.chiefscientist.qld.gov.au/publications/understanding-floods/flood-consequences (Retrieved 12th January, 2018)

12. Rather I.A., Parray H.A., Lone J.B., Paek W.K., Lim J., Bajpai V.K. and Park Y.H. 2017. Prevention and control strategies to counter dengue virus infection. *Frontiers in cellular and infection microbiology* 336: 1-8

13. Relief Web 2017. Sri Lanka's deadly floods could exacerbate dengue crisis, warns Save the Children as more rainfall is on the way.

https://reliefweb.int/report/sri-lanka/sri-lanka-s-deadly-floods-could-exacerbate-dengue-crisis-warns-save-children-more (Retrieved 12th January, 2018)

14. Sri Lanka Disaster Knowledge Network 2009. Hazard profile.

    http://www.saarc-sadkn.org/countries/srilanka/hazard_profile.aspx (Retrieved on 12th January, 2018)

15. The Guardian 2017. Floods and landslides in Sri Lanka kill at least 150 people

    https://www.theguardian.com/world/2017/may/28/floods-and-landslides-and-in-sri-lanka-kill-at-least-150-people (Retrieved 15th January, 2018)

16. UNDP 2017. UNDP responding to Tropical Cyclone Mora in Sri Lanka, Bangladesh, Myanmar and India.

    http://www.undp.org/content/undp/en/home/presscenter/pressreleases/2017/05/31/undp-responding-to-tropical-cyclone-mora-in-sri-lanka-bangladesh-myanmar-and-india/ (Retrieved 12th January, 2018)

17. WHO 2017. Dengue fever – Sri Lanka.

    http://www.who.int/csr/don/19-july-2017-dengue-sri-lanka/en/ (Retrieved 16th January, 2018)

18. WHO 2018a. Dengue control.

    http://www.who.int/denguecontrol/control_strategies/environmental_management/en/ (Retrieved 16th January, 2018)

19. WHO 2018b. Flooding and communicable diseases fact sheet, Humanitarian Health Action.

20. WHO 2018c.Health and development.

    http://www.who.int/hdp/en/ (Retrieved 15th January, 2018)

Chapter 5

Effect of Flooding in Kaluwanchikudy DSD: An Analysis Using GIS Application

S. Mathanraj[1]* *and M.I.M. Kaleel*[2]

[1]*Eastern University, Sri Lanka, Department of Geography, Vantharumoolai, Sri Lanka*

[2]*South Eastern University of Sri Lanka, Department of Geography, Oluvil, Sri Lanka*

**E-mail: mathanrajs@esn.ac.lk*

ABSTRACT

Flooding is the natural hazard that poses the greatest concern in the coastal region of Sri Lanka. The major objective of this study is to identify the flooding severity levels in the 52.5 km² Kaluwanchikudy Divisional Secretariat using a GIS application. Primary data are personal observations, and face-to-face interviews. Secondary data are Disaster Management Center and Census reports, images and published research reports. The severity level of flooding in this area was analyzed using the Shuttle Radar Topography Mission (SRTM) image in ArcGIS overlaid on Google Earth Pro map making tool. It was found that about half the area is affected by flooding, with the highly, moderately and slightly affected areas all about 9 km² in extent. The major causes of flooding are poor drainage systems in some region, low land, unclean drainage systems, and up-and-down roads. The map can be used to evacuate people from the worst affected areas during the flood season, or to guide resettlement in the highland area.

Keywords: *Disaster management, SRTM, Severity level, Flooding, Drainage system.*

1. Introduction

Water is the most precious resource to human life. However, this valuable resource may cause harm and loss when it is implicated in a flooding disaster. Flooding is the severe flow of water towards low-lying land as a result of continuous rainfall. Flooding may also result from the overflow from rivers and lakes, in which the water overtops or breaks levees and escapes its usual boundaries (Awate, 2016). It is one of the most common natural disasters in the study area.

Floods often cause damage to homes and businesses if they are in the natural flood plains of rivers. Some floods develop slowly while others, such as flash floods, can develop in just a few minutes and without visible signs of rain (Ilangovan, 2009). Additionally, floods can be local, impacting a neighborhood or community or very large, affecting entire river basins.

Planners identify and analyze options for reducing potential flood losses and protecting natural values. By this, they identify the factors that most affect the degree of hazard, the extent of development that is at risk, and the potential components of hazard mitigation programme (Dennis and Joanne, 2006).

1.1. Study Area (Figure 1.1)

The District of Batticaloa consists of several administrative divisions. One of these, the Kaluwanchikudy Divisional Secretariat Division, is located in the south eastern part of the district. It has an extent of 52.5 km^2 and consists of 45 Grama Niladhari Divisions which is the subunit of a divisional secretariat and contains 136 villages. Its population is 63,108 consisting of 17,784 families. Their main occupations are farming and fishing. Residents in 30 Grama Niladhari Divisions were badly affected by the 2004 tsunami (Savithri and Priyanga, 2006).

1.2 Several Types of Flood occur in the Study Area

Areal (rainfall-related): Floods can happen on flat or low-lying areas when the ground is saturated and water either cannot run off or cannot run off quickly enough to stop accumulating. This may be followed by a river flood as water moves away from the floodplain into local rivers and streams (Awate, 2016).

Riverine: River flows may rise to flood levels at different rates, from a few minutes to several weeks, depending on the type of river and the source of the increased flow (Awate, 2016).

Slow-rising floods: Most commonly occur in large rivers with large catchment areas in the world. The increase in flow may be the result of sustained rainfall, monsoons such as Southwest and Northeast, or tropical cyclones as Hurricane intensity in the Bay of Bengal North Indian Ocean cyclone). Localized flooding may be caused or exacerbated by drainage obstructions such as landslides (Awate, 2016).

The main objectives of this study are: (i) to identify the flood severity zones, (ii) to create a flooding map to the study area, and (iii) to predict the effects of flooding.

2. Methodology

Primary Data

Direct personal observation in the flooded zone and face-to-face interviews with two Disaster Management officials, eight Grama Niladhari and 30 residents.

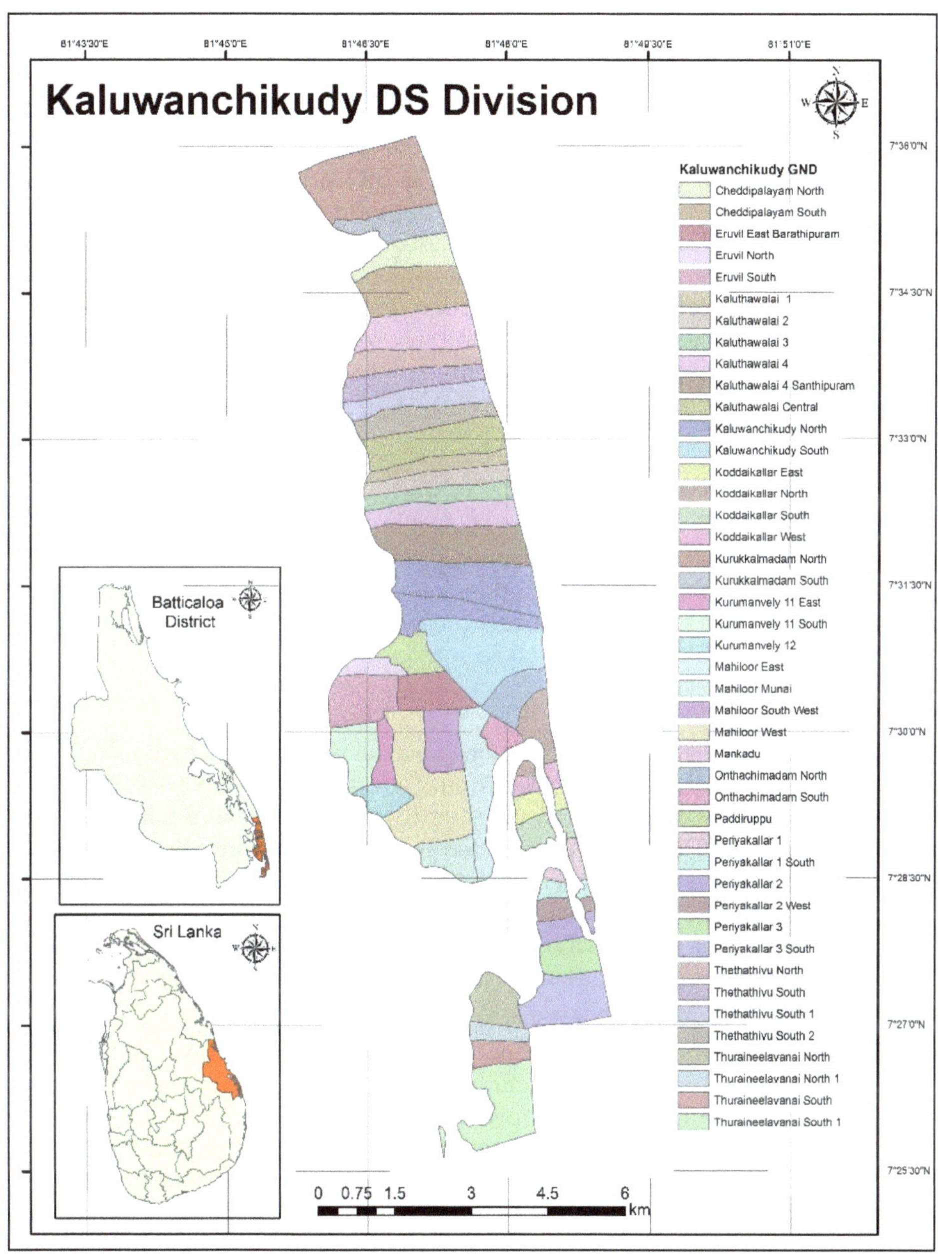

Figure 5.1: The Study Area.

Secondary Data

Disaster Management Center reports, census reports, rainfall and temperature data from meteorological department, various images and published research reports, and the ***Shuttle Radar Topography Mission*** **(*SRTM*)** image from Earth Explorer.

Data Analysis

The SRTM image was used to examine the severity level of flooding in this area and to define the flooding severity zones. The SRTM image of Kaluwanchikudy area has downloaded from Earth explorer that used after removing noise. Water body of this SRTM raster image was converted to the vector polygon for finding the flood prone area. The sensitivity zone was bordered as high, moderate and low sensitivity area and the high zone bordered below 100m from the water body, moderate zone between 100m to 200m and low sensitivity zone between 200m to 300m. ArcGIS 10.4.1 software was utilized to prepare the map.

3. Results

3.1 Flooding Severity Levels

Three severity levels were defined based on distance from the water bodies. Areas within 100 m of water bodies were defined as high-severity zones (8.95 km^2); areas between 100m and 200 m as moderate-severity zones (9.19 km^2); and areas between 200 m and 300m as low-severity zones (9.31 km^2). The flooding severity levels in Kaluwanchikudy Divisional Secretariat Division are shown in **Figure 5.2**.

3.2 Cause of Floods

The study area is located in the monsoonal climatic zone, which usually gets its rainfall in the northeast monsoon period. The annual rainfall varies from 864 to 3081mm with the evidence of past years in this area. This severe rainfall directly caused to the flood. This is a low land often got the sudden water flow cause to the flooding and also this area has very less drainage facilities. Further, this area does not contain the proper drainage facilities and very rare maintenance of drainage because of developing region is mostly the reason for flooding. In addition, wet lands area nowadays changing as settlement land. Lots of mangrove plants destroyed by the flood events in past years.

In summary, the causes of floods were found to be both natural and anthropogenic. Natural factors includes heavy rainfall, topography such low land, sudden rainfall, monsoons, low drainage patterns, lack of vegetation or woodland by the effects of natural disaster and human activities, and little to slow the floodwater down. Manmade factors include drains blocked by waste, improper drainage facilities, lack of drainage facilities, reduction of wet lands, and damage of drainage, improper infrastructure lack of wet lands by buildings and improper disposal of solid wastes (Mathbor 2007, Shannon *et al.*, 2013).

According to the **Figure 5.3**, the major causes for facing the flood disaster in this area is lack of drainage system 34 per cent in some flooded area, low land 31 per cent, and unclean drainages 23 per cent, and ups and downs roads 12 per cent (Field data, 2017).

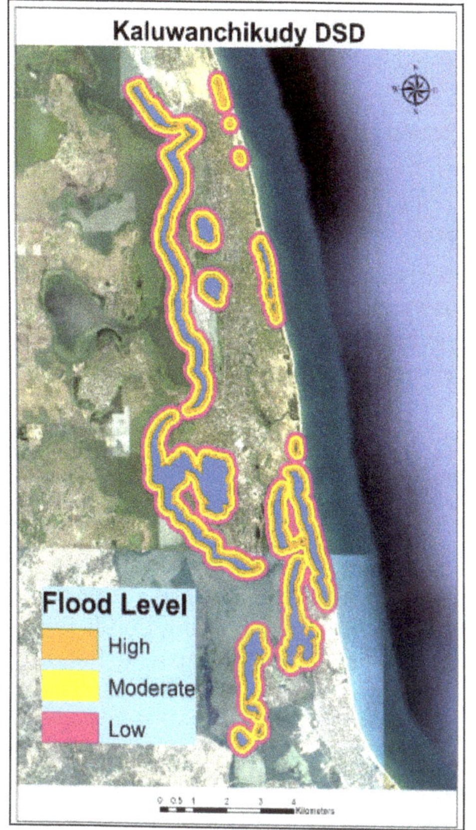

Figure 5.2: Flood Affected Zone.

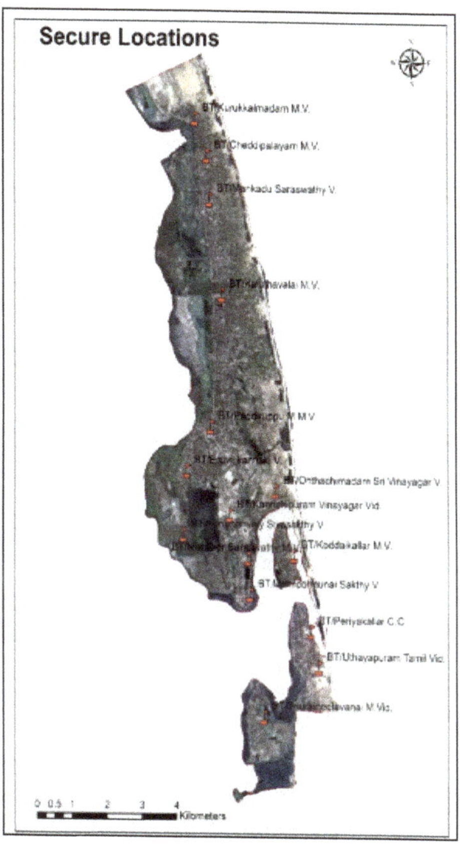

Figure 5.3: Safe Zones.

3.3 Effects of Flood in Study Area

The primary effects were found to be:

- Damage to buildings and other structures, including roadways and drainage.
- Loss of drinking water treatment facilities, which may result in loss of drinking water or severe water contamination.
- Damage to houses, property and important possessions such as furniture, other electrical appliances.
- Inundation of farm land, making the land unworkable and preventing crops from being planted or harvested, and affect the livestock. This can lead to shortage of food both for humans and farm animals.
- People forced to move from their house to temporary area as a refuge.
- Lack of hygienic foods

Secondary and long-term effects were found to be:

- Psychological damage to those affected, in particular where serious injuries and loss of property occurred.
- Small businesses that never reopened their doors following a flooding disaster

4. Conclusions Recommendations

The study area is often affected by flood disasters. According to the interviews, every year those people are facing the flood event. The flood-severity map can be used by the Disaster Management Centre to evacuate people from the worst affected areas during the flood season, and to guide resettlement in the highland area.

5. Recommendations

There are several ways to control flooding that should be considered by the local authorities.

Diversion Canals

Floods can be controlled by redirecting excess water to purpose-built canals or floodways, which in turn divert the water to temporary holding ponds in low land area or other bodies of water where there is a lower risk or impact to flooding.

Self-Closing Flood Barrier

The *self-closing flood barrier* (SCFB) is a flood defense system designed to protect people and property from inland waterway floods caused by heavy rainfall. The SCFB can be built to protect residential properties and whole communities, as well as industrial or other strategic areas. The barrier system is constantly ready to deploy in a flood situation, it can be installed in any length and uses the rising flood water to deploy.

Should a flood be anticipated or occur, there are actions that should be taken:

Before the Floods

- Know about your local relief centers and evacuation routes.
- Keep emergency numbers and important information handy as well as emergency supplies, kits, first aid items. These may include water, canned food, can opener, battery-operated radio, flashlight and protective clothing.
- Fold and roll up anything onto higher ground (or upper floors of your home) including chemicals and medicines.
- Make sure everything that is of importance secured (jewellery, documents, pets and other valuables).
- Keep a lot of vegetation in your compound including Plant trees and shrubs, if you are living in a low-lying area. This may control erosion and help soften the speed of the flowing water.

During the Flood

- ☆ Listen to a battery-operated radio for the latest information.
- ☆ Turn off all utilities at the main power switch and close the main gas valve if advised to do so.
- ☆ If you have come in contact with floodwaters, wash your hands with soap and disinfected water.
- ☆ If flooding occurs, go to higher ground and avoid areas subject to flooding.
- ☆ Do not attempt to walk across flowing streams or drive through flooded roadways.
- ☆ If water rises in your home before you evacuate, go to the top floor or roof.
- ☆ Move valuables, important papers and clothing to upper floors.
- ☆ If you have only one floor, put items on shelves, tables or countertops.
- ☆ Sanitize your bathtub and sinks and fill them with fresh, clean water in case the water supply becomes contaminated.
- ☆ If you feel threatened by rising water, leave your home or move to upper floors.
- ☆ Stay away from downed power lines.

After the Flood

- ☆ If it's dark, use a flashlight – not matches, a candle or a lighter.
- ☆ Listen for reports to see when drinking water is safe again.
- ☆ Begin initial cleanup as soon as waters recede. Separate damaged from undamaged items, clean and disinfect everything that got wet.
- ☆ Check for structural damage before re-entering your home to avoid being trapped in a building collapse.
- ☆ Take photos of any floodwater in your home and save any damaged personal property.
- ☆ Keep power off until an electrician has inspected your system for safety.
- ☆ Boil water for drinking and food preparation until authorities tell you that your water supply is safe.
- ☆ When cleaning, wear a mask, gloves and cover all to minimize exposure to possible hazardous materials.
- ☆ Wet items should be cleaned with a pine-oil cleanser and bleach, completely dried, and monitored for several days for any fungal growth and odors.

References

1. Awate S.J. 2016. Environmental Geography. Luxmi Book Publications, India, pp 52-54

2. Ilangovan P. S. 2009. Perception of flood risk: a case study from the flood affected areas of Madhurai. *Natural Hazards and Disaster Essay on Impact and Management*. International Conference Volume. Anantapura: Department of Geography, Sri Hrisna Devaraya University, India.

3. Dennis M. and Joanne M.N. 2006. Natural hazards and disasters: Drawing on the international experiences from disaster reduction in developing countries. A Report by Norwegian Institute for Urban and Regional Research (NIBR), Jan. 16th 2006.

4. Field Data from Earth explorer 2017.

   Available from: https://earthexplorer.usgs.gov/, Accessed: 2017/10/30

5. Mathbor G.M. 2007. Enhancement of community preparedness for natural disasters: The role of social work in building social capital for sustainable disaster relief and management. *International Social Works* 50: 357-369

6. Savithri K. and Priyanga A. 2006. Yaalpana Nahara Vella Anartha Muhamaituvam, Yaalpana Puviyiyalalan, Volume 20, Jaffna University, Jaffna, Sri Lanka.

7. Shannon Doocy, Amy Daniels, Sarah Murray and Thomas D. Kirsch, 2013. The Human Impact of Floods: a Historical Review of Events 1980-2009 and Systematic Literature Review. *PLOS Currents Disasters* Apr 16. Edition 1. doi: 10.1371/currents.dis.f4deb457904936b07c09daa98ee8171a.

Chapter 6

Flood Susceptibility Mapping Using the Analytical Hierarchy Process Method and Geographic Information System: Application to the Savannah Region, Togo

K. Komi

West African Science Service Center on Climate Change and Adapted Land Use (WASCAL), University of Lomé, Lomé, Togo
E-mail: kossik81@yahoo.fr

ABSTRACT

Flood damages have increased in the last two decades in West Africa. Flood hazard mapping is an important component for efficient flood risk management. The aim of this study is to assess physical vulnerability to flood at the regional scale in Togo. The methodology used in this research is based on the integration of remote sensing data including rainfall, elevation, drainage density, soil (infiltration), land cover/use and slope within a Geography Information System (GIS). The relative weight of each parameter for the flood hazard is computed using an "Analytical Hierarchy Process" (AHP). The results show that areas with high and very high level of physical vulnerability to flood covers 32 per cent and 2 per cent of the study area, respectively. The proposed map can be used to identify areas vulnerable to flood, to promote greater awareness about the risk of flood, and to prioritize the flood mitigation efforts in the country.

Keywords: *Flood vulnerability, Analytical hierarchy process, Geographic information system.*

Abbreviations

ICSU: International Council for Science Union
GIZ: German International Cooperation Service
OECD: Organization for Economic Cooperation and Development
SRTM: Shuttle Radar Topographic Mission

1. Introduction

Floods are the most damaging natural hazards in the world. Floods in Europe cost about US$ 4.2 billion per year from the periods 2000 to 2012 (Jongman *et al.*, 2014). About 20 per cent of the world's population could be affected by floods in 2050, with economic damage that could rise to US$ 45 trillion without taking the impacts of climate change on flood risk into account (OECD, 2012). From 1900 to 2006, floods in Africa have killed nearly 20,000 people and caused damages of about US$ 4 billion (ICSU, 2007). Heavy rainfall in September 2007 caused the worst floods that West Africa had faced in many decades. The hardest hit countries were Togo, Burkina Faso and Ghana with 23, 46 and 56 persons killed respectively (Tschakert *et al.*, 2010). Furthermore, in 2008, severe floods killed 20 people and destroyed more than one hundred bridges in Togo (Amegadje, 2009). More recently, 38 people were reported killed and 9,000 homes were destroyed in the August-September 2016 floods of the Niger River. These high tolls highlighted major gaps in the management of flood risk in West Africa. Thus, flood susceptibility maps are important as they provide the basis for the implementation of flood risk management strategies.

Many methods have been used to identify areas vulnerable to flood. For instance, Wang *et al.* (2011) used multi-criteria analysis to estimate flood vulnerable areas while Kourgialas and Karatzas (2011) estimated flood hazardous areas by overlaying GIS-layers which visualize spatial and climate information. In Komi *et al.* (2017), the authors used both hydrological and hydraulic models to simulate flood hazard extent. Brimicombe and Barlett (1996) coupled a hydraulic model with a GIS and a Digital Elevation Model (DEM) to map the areas and depths of inundation.

The present paper applied a GIS-based multi-criteria analysis in the Savannah region in Northern Togo. The aim is to identify areas susceptible to flood where mitigation measures should be taken. Thus, a ***Flood Susceptibility Index* (*FSI*)** has been introduced to define these areas.

2. Materials and Methods

2.1. Description of the Study Area

The study area is located in the northern Togo from Latitude 9° 52′48″ N to 11° 7′ 48″N and Longitude 0° 8′24″W to 0° 54′36″E and has an area of 8,470 km². The population in 2010 was about 828,200. The main economic activities are agriculture and livestock. The maximum elevation of this area is 512 m, while the minimum elevation is 104 m (**Figure 6.1**).

The climate of this area is tropical semi-arid and is characterized by a rainy season (April to October) and a dry season (November to March). The mean annual

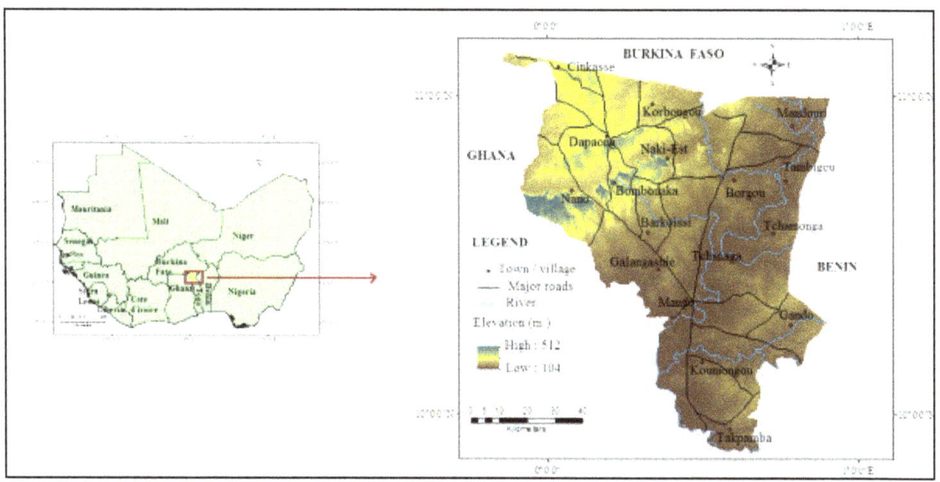

Figure 6.1: Location of the Study Area.

rainfall at Mango gauge station was 1,047 mm from 1980 to 2012. The maximum daily rainfall observed at Mango during the period 1980-2012 ranged from 48.9 mm in 2001 to 132.7 mm in 1999 (**Figure 6.2a**). August is the rainiest month in the study area while March and April are the hottest months (**Figure 6.2b**).

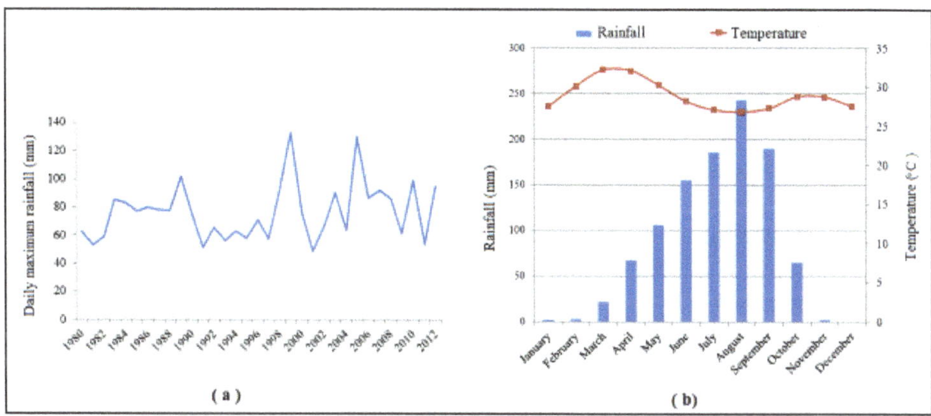

Figure 6.2: The Month-wise Rain Data Recorded at Mango Gauge Station, (a) Daily Maximum Rainfall; (b) Climate Diagram at Mango.

2.2. Methods

In this study, the author has integrated remote sensing data, using a Geographic Information System (GIS) and the Analytical Hierarchy Process (AHP) to generate flood susceptibility map (**Figure 6.3**). Using a GIS, thematic maps of five causative factors of flood susceptibility were prepared, *viz.* Drainage Density (DD), Elevation (E), Land Use (LU), Rainfall (R), Slope (SL) and Soil Infiltration Rate (SIR). The selection of these parameters was based on their relevance to flood hazards (Kazakis

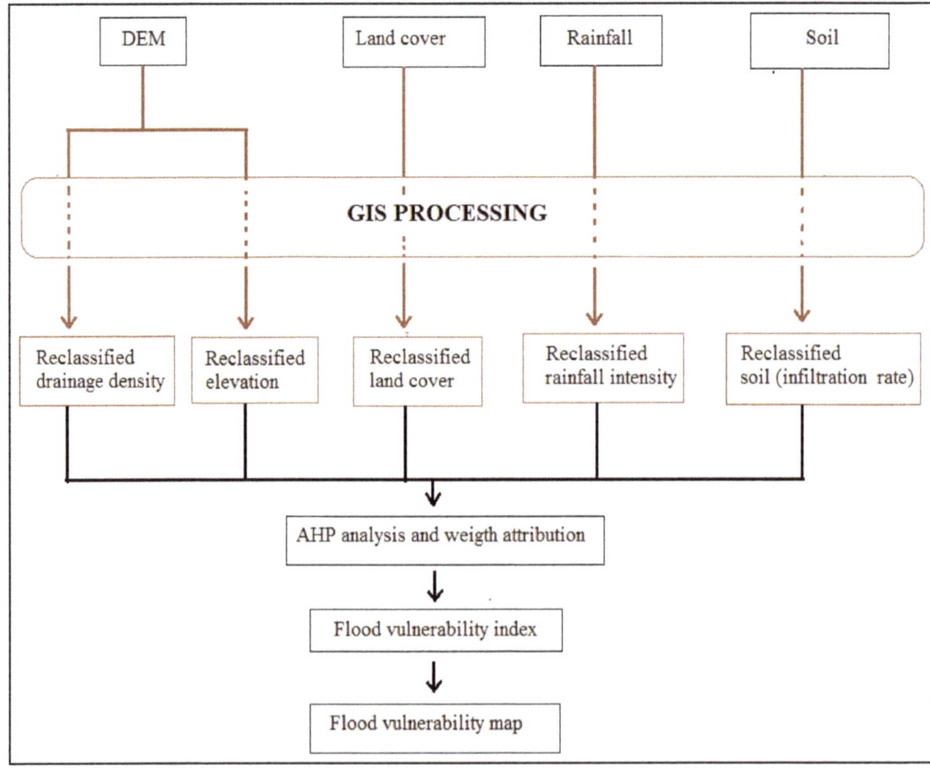

Figure 6.3: Flowchart of the Methodology.

et al., 2015, Ouma and Tateishi 2014). The DD was generated from the Shuttle Radar Topographic Mission (SRTM) and Digital Elevation Model (DEM) constructed at 30 m spatial resolution by computing the ratio of the total length of streams in a catchment over its contributing area. Land use information of 2015 was collected from German International Cooperation Service (GIZ) Togo, while the rainfall intensity was expressed as the total rainfall of August 2015 at each grid in the study area and estimated from United States Geological Survey (USGS) data. Moreover, the slope (as percentage) was extracted from the DEM. The soil map was obtained from the Harmonized World Soil Data. The Soil Infiltration Rate (SIR), used in this study as flood causative factor was obtained from Johnson (1991) and Kern (2010).

2.2.1 Flood Susceptibility Index (FSI)

In the present research, an index model was developed in a GIS in order to help in the identification of hotspots related to flood susceptibility at a regional scale. The FSI was calculated using *Equation 1*:

$$FSI = S_{DD} * W_{DD} + S_E * S_{LU} * W_{LU} + S_R * W_R + S_{SL} * W_{SL} + S_{SIR} * W_{SIR} \quad [1]$$

Where S_i is a ranking score and W_i is the weight of the parameter *i*. The scores were obtained by assigning each grid point a scale 1, 2, 3, 4 or 5 based on their importance in causing flood. The classes of the drainage density, elevation, rainfall intensity were defined using the natural breaks method of classification, similarly to previous studies (Kazakis *et al.*, 2015, Huan *et al.*, 2012). The land use classes where estimated from Rimba *et al.* (2017), whereas the classes of slope and soil infiltration rate were estimated from Demek (1972) and Brouwer (1990) respectively. The weights of the parameters indicate their relative importance in contributing to flood susceptibility. They were computed using the AHP method (Saaty 1977). Furthermore, the FSI was classified into five susceptibility classes "very low", "low", "moderate", "high" and "very high" with the indices 1, 2, 3, 4 and 5, respectively.

2.2.2 Analytical Hierarchy Process (AHP)

AHP is a multi-criteria, mathematically based method which uses a set of pair-wise comparison matrices to estimate the relative importance of different criteria and alternatives among which the best decision is made. The relative importance between the criteria is assessed from 1 to 9 showing less significant to more significant respectively (**Table 6.1**). The Saaty's AHP model has attracted the interest of many researchers (Omkarprasad and Sushil 2006, Bathrellos *et al.*, 2016 and Komi *et al.*, 2016) because it has the advantage of incorporating a test for checking the consistency of a choice, thus reducing the uncertainty in the evaluation process.

Table 6.1: Importance Scale Used in the Pair-wise Comparison Matrix

Intensity of Importance	*Definition*	*Description*
1	Equal importance	Two criteria contribute equally to the objective
3	Moderate importance	Experience and judgment slightly favor one over the other
5	Strong importance	Experience and judgment strongly favor one over the other
7	Very strong importance	Experience and judgment very strongly favor one criterion over the other. Its importance is demonstrated in practice
9	Extreme importance	The evidence favoring one criterion over the other is of the highest possible validity
2, 4,6, 8	Intermediate values	When agreement is necessary

In order to compute the weights for each parameter, the AHP starts creating a pair-wise comparison matrix $M = (B_{ij})$. Each numerical value B_{ij} of M represents the relative importance of the ith parameter in comparison with the jth parameter. If $B_{ij} > 1$, then the ith parameter is more important than the jth parameter, whereas if $B_{ij} < 1$, then the ith parameter is less important than the jth parameter. If two parameters have the same importance, then $B_{ij} = 1$. The numerical values satisfy the condition given in *Equation 2*.

$$B_{ij} * B_{ji} = 1 \quad [2]$$

After building the matrix M, a normalized pair wise comparison matrix was derived by dividing each value B_{ij} by the sum of all values of that column. Finally, the relative weight (W_t) vector was estimated by averaging the values on each row of the normalized pair-wise comparison matrix. The AHP method requires all indicator weights to satisfy the condition shown in *Equation 3.*

$$\sum_{i=1}^{n} W_t = 1 \quad [3]$$

The AHP method provides the possibility to check the consistency of the estimated weights. This is done with the ***Consistency Ratio*** **(*CR*)** which is shown in *Equation 4.*

$$CR = \frac{CI}{RI} \quad [4]$$

Where, ***Consistency Index*** **(*CI*)** is obtained by first computing the scalar λ_{max} as the average of the elements of the vector whose *i*th element is the ratio of the *i*th element of the vector $(M * W_t)$ to the corresponding element of the vector W_t.

Then, CI is calculated using the *Equation 5.*

$$CI = \frac{\lambda_{max} - n}{n - 1} \quad [5]$$

Where λ_{max} the largest Eigen value of the matrix and n is is the number of parameters. The ***Random Index*** **(*RI*)** is a constant which depends on *n* (**Table 6.2**). When CR < 0.1, the evaluation is consistent and reliable results can be expected from the AHP model.

Table 6.2: Random Index (RI)

n	1	2	3	4	5	6	7	8	9	10	11	12	13
RI	0	0	0.58	0.90	1.12	1.24	1.32	1.41	1.45	1.49	1.51	1.48	1.56

3. Results and Discussion

3.1 Scores of the Flood Susceptibility Parameters

The classes of the six parameters and the corresponding scores are shown in **Table 6.3.**

Table 6.3: Classes of the Parameters and the Corresponding Scores

Parameters	*Classes*	*Score*
Drainage density (m/m^2)	0-0.0028	1
	0.0028-0.0087	2
	0.0087-0.0181	3
	0.0181-0.0371	4
	0.0371-0.1259	5

Parameters	*Classes*	*Score*
Elevation (m)	103-154.58	5
	154.58-206.15	4
	206.15-267.4	3
	267.4-354.43	2
	354.43-514	1
Land use/cover	Wetland	5
	Built-up land	4
	Agriculture	3
	Sparse vegetation	2
	Forest	1
Rainfall intensity (mm)	224-236	1
	236-250	2
	250-259	3
	259-271	4
	271-308	5
Slope (per cent)	0-2	5
	2-5	4
	5-15	3
	15-35	2
	35-60	1
Soil infiltration rate (mm/h)	1-5	5
	5-10	4
	10-20	3
	20-30	2
	>30	1

The drainage density values vary in a range between 0 and 0.1259 with the highest values occurring downstream of Mandouri (**Figure 6.4a**). Other factors being equal, higher values of drainage density indicate higher susceptibility to flood. In the study area, the highest elevation appears in the northwestern part, while the lowest elevation represents about 20 per cent of the region area (**Figure 6.4b**).

A large proportion of this region has a slope comprised between 0 and 5 per cent (**Figure 6.4e**). Both low elevation and low slope are indicative of flood prone areas. They were attributed the highest score. Moreover, the study area is predominantly covered by sparse vegetation, which was assigned a score equal to 2 (**Figure 6.4c**). The total rainfall in August 2015 ranges from 224 to 308 with the highest value located in the northern part of the study area (**Figure 6.4d**). Also, the rate of water infiltrating the soil is considered as a flood susceptibility parameter. It is closely related to the texture of the soil. The lowest infiltration rate (highest score) is located in the upper left corner of the study area (**Figure 6.4f**).

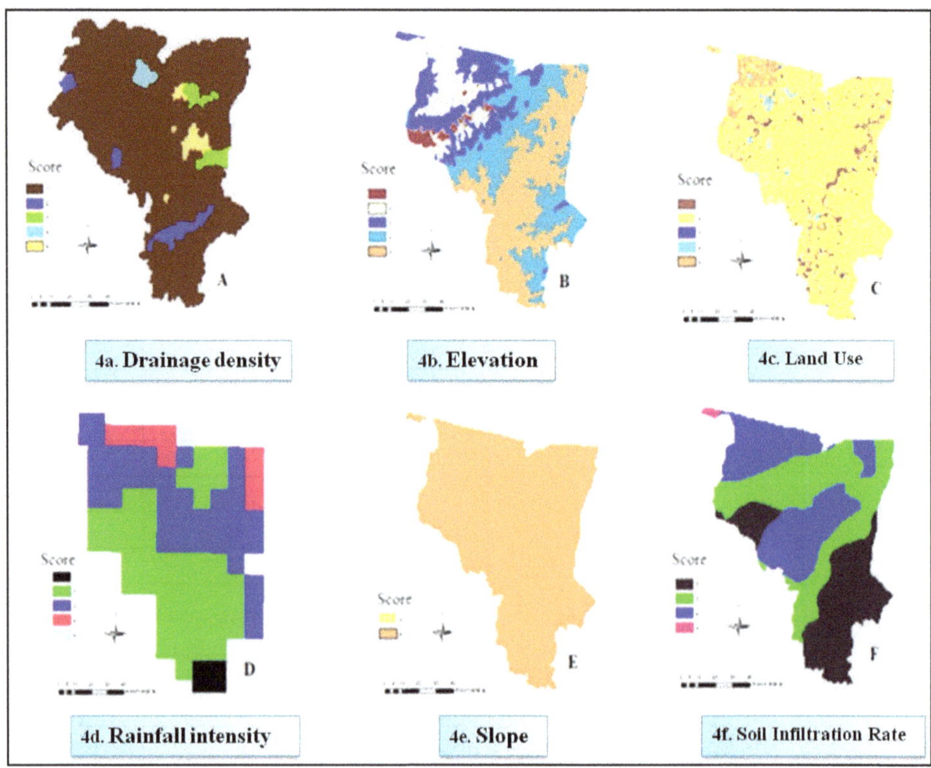

Figure 6.4: Thematic Maps of the Parameters Used in this Study.

3.2. Relative Weights of the Parameters

The results of the pair-wise comparison and the corresponding relative weights are presented in **Table 6.4**.

Table 6.4: Pair-wise Comparison Matrix

Parameter	*Drainage Density*	*Elevation*	*Land use*	*Rainfall Intensity*	*Slope*	*Soil*	W_i
Drainage density	1	1	2	3	4	5	0.25
Elevation	1/2	1	3	4	5	6	0.38
Land use	1/2	1/3	1	2	3	4	0.16
Rainfall intensity	1/3	1/4	1/2	1	2	3	0.10
Slope	1/4	1/5	1/3	1/2	1	2	0.07
Soil	1/5	1/6	1/4	1/3	1/2	1	0.04
λ_{max} **= 6.13 n =6 RI= 1.24 CI= 0.03 CR= 0.02**							

Since the obtained CR's value is much lower than the threshold value of 0.1, the pair-wise matrix is consistent and the determined weights are acceptable.

These weights indicate the importance of each parameter as compared to the other parameter in contributing to physical flood vulnerability. Elevation has been considered the most important parameter because flood areas are usually located in low elevation and low slope. However, the slope has lower importance in this study because it is somehow considered in the elevation parameter (Kazakis *et al.*, 2015). The drainage density was considered as the second more important parameter although this parameter was not prioritized in other studies (Danumah *et al.*, 2016, Ouma and Tateishi 2014) because flood prone areas are often located near drainage networks. Land use and rainfall intensity are at the third level of importance in alignment to similar study (Kazakis *et al.*, 2015). The soil infiltration rate can be of high importance for flood generation, particularly in urban areas. As this is not the case for our study area which is predominantly rural, soil water infiltration rate has been attributed the lowest weight.

4.2. Flood Susceptibility Map

After estimating the relative weights of each parameter, the author performed a multi-criteria analysis to compute FSI using *Equation 1* and produced a flood susceptibility map in a GIS environment (**Figure 6.5**). The results show that 32 and

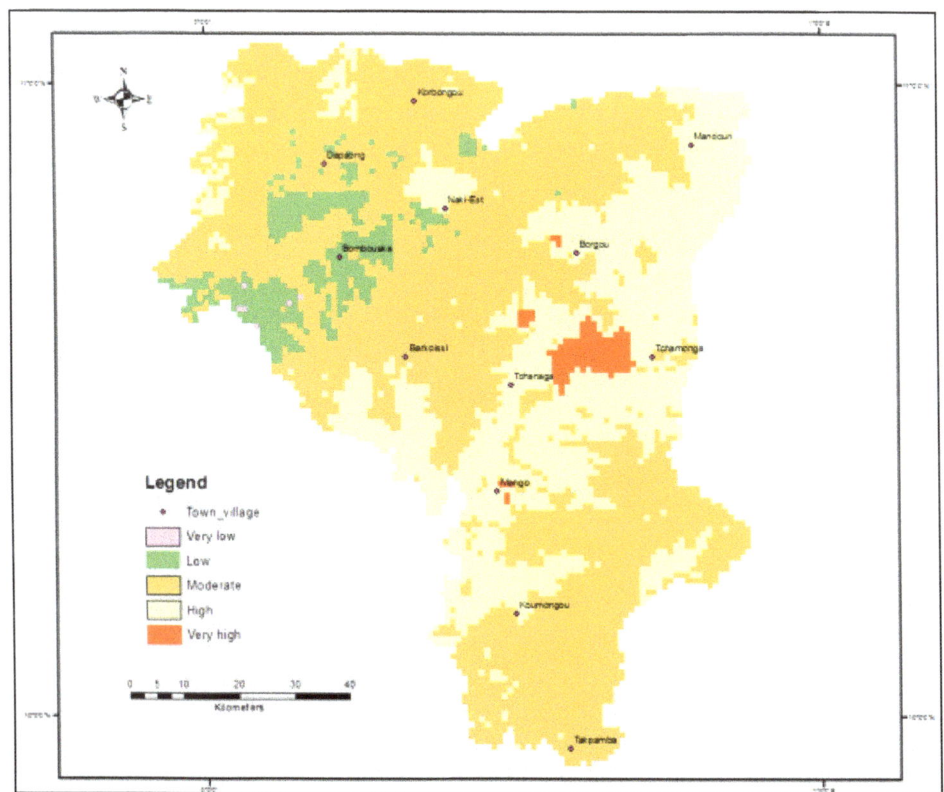

Figure 6.5: Flood Susceptibility Map of the Study Area.

2 per cent of the study area fall in the high and very high classes respectively. These areas are those in low land areas and close to the Oti River. In addition, 61 per cent of the study area was classified as moderate level of susceptibility to flood, while nearly 6 per cent falls in the low class. Less than 1 per cent is in the very low class. This area located at a higher elevation and far from a drainage network.

4. Conclusions

The main objective of this study is to identify areas susceptible to flood in the Savannah region of Togo. This is important for mitigation measures because it provides an assessment of flood vulnerable areas in the study area. The proposed methodology analyses spatially six parameters (drainage density, elevation, land use, rainfall intensity, and slope and soil water infiltration rate). The relative importance of each parameter was calculated using the AHP method. Thematic maps of these parameters were combined in a GIS environment to create physical flood vulnerability map. A sensitivity analysis, which evaluates the effects of each parameter on the overall physical flood vulnerability index, will be performed in future work.

Acknowledgements

The author is grateful to Prof. Kouami Kokou and Kossiwa Zinsou-Klassou for their scientific assistance. I would like to thank the Centre for Science and Technology of the Non-Aligned and Other Developing Countries (NAM S&T Centre) for funding my participation to this roundtable as well as National Science and Technology Commission (NASTEC) for the hospitality in Sri Lanka.

References

1. Amegadje M.K. 2009. Profil environnemenal du Togo: rapport provisoire. RepubliqueTogolaise, 124p.
2. Bathrellos G.D., Karymbalis E., Skilodimou H.D., Gaki-Papanastassiou K. and Baltas, E.A. 2016. Urban flood hazard assessment in the basin of Athens Metropolitan city. *Greece. Environ. Earth. Sci.* 75: 319
3. Brimicombe A.J. and Barlett J.M. 1996. Linking geographic information systems which hydraulic simulation modelling for flood risk assessment: The Hong Kong Approach; Goodchild M.F., Ed; GIS and Environmental modelling, Oxford University Press: New York, NY, USA, pp.113-140.
4. Brouwer C. 1990. Gestion des Eaux en Irrigation. Manuel de formation No5, Methodes d'irrigation. Division de la mise en valeur des terres et des eaux. Food and Agriculture Organization (FAO): Rome, Italy (*In French*)
5. Danumah H. J., Odai N. S., Saley M., B., Szarzynski J., Thiel M., Kwaku A., Kouame K.F., Akpa Y.L. 2016. Flood risk assessment and mapping in Abidjan district using multi-criteria analysis (AHP) model and geoinformation techniques (Cote d'Ivoire). *Geoenvironmental Disasters* 3: 10

6. Demek J. 1972. Manual of Detailed Geomorphological Mapping. International Geographical Union. Commission on Geomorphological Survey and Mapping. Prague, Academia, 1972

7. Huan H., Wang J., Teng Y. 2012. Assessment and validation of groundwater vulnerability to nitrate based on modified drastic model: a case study in Jilin City of northeast China. *Sci. Total Environ* 440:14-23

8. ICSU 2007. Draft science/work plan. Natural and Human-induced Hazards and Disasters in sub-Saharan Africa, Report tabled at the Regional Consultative Forum, Boksburg, 25-27 September 2006

9. Johnson A.I. 1991. A field method for measurement of infiltration. 2nd ed., U.S. Geological Survey, Denver, CO, USA

10. Jongman B., Hochrainer-Stigler S., Feyen L., Aerts J.C. J. H., Mechler R., Botzen W.J.W., Bouwer L.M., Pflug G., Rojas R., Ward P.J. 2014. Increasing stress on disaster-risk finance due to large floods. *Nature Climate Change* 4: 264-268

11. Kazakis N., Kougias I., Patsialis T 2015. Assessment of flood hazard areas at a regional scale using an index-based approach and Analytical Hierarchy Process: Application in Rhodope-Evros region, Greece. *Science of the Total Environment*, 538: 555-563

12. Kern J. 2010. Water use and conservation. California Department of Water Resources: Sacramento Country, CA, USA.

13. Komi K., Amisigo A.B. and Diekkrüger B. 2016. Integrated flood risk assessment of rural communities in the Oti River Basin. *Hydrology* 3: 42

14. Komi K., Neal J., Trigg A.M. and Diekkrüger B. 2017. Modelling of flood hazard extent in data sparse areas: a case study of the Oti River basin, West Africa. *Journal of hydrology: regional studies* 10: 122-132

15. Kourgialas N.N. and Karatzas G.P. 2011. Flood management and a GIS modeling method to assess flood-hazard areas: a case study. *Hydrol Sci J* 56: 212-225

16. OECD 2012. OECD Environmental Outlook to 2050: the Consequences of Inaction. OECD Publishing, Paris, France.

17. Omkarprasad V. and Sushil K. 2006. Analytic hierarchy process: an overview of applications. *European Journal of Operational Research* 169:1-29

18. Ouma Y., Tateishi R. 2014. Urban flood vulnerability and risk mapping using integrated multi-parametric AHP and GIS: methodological overview and case study assessment. *Water* 6: 1515-1545

19. Rimba B.A., Setiawati D.M., Sambah B.A., Miura F. 2017. Physical flood vulnerability mapping applying geospatial techniques in Okazaki City, Aichi Prefecture, Japan., Urban Sc., 1,7.

20. Saaty T.L. 1977. A scaling method for priorities in hierarchical structures. *J Math Psychol* 15: 234-281

21. Tschakert P., Sagoe R., Ofori-Darko G., and Codjoe S.N. 2010. Floods in the Sahel: an analysis of anomalies, memory, and anticipatory learning. *Climatic Change* 103: 471–502.

22. Wang Y., Li Z., Tang Z., Zeng G. 2011. A GIS-based spatial multi-criteria approach for flood risk assessment in the Dongting Lake Region, Hunan, Central China. *Water Resour. Manag* 25: 3465-3484

    https://www.unisdr.org/2015/docs/climatechange/COP21_WeatherDisastersReport_2015_FINAL.pdf (accessed on Nov. 19, 2017).

Chapter 7

Applying Robust Decision Making (RDM) to Ensure Robust Flood Management in Ho Chi Minh City, Vietnam

B.T. Sinh

National Institute of S&T Policy and Strategy Studies (NISTPASS),
38 Ngo Quyen Str, Ha Noi, Vietnam,
E-mail: btsinh@most.gov.vn

ABSTRACT

Ho Chi Minh City (HCMC) faces significant flood risk driven by hard-to-predict future climate and socio-economic trends. Many regions in the world face similar risk management challenges. However, traditional quantitative methods for risk management often prove inadequate because they may produce strategies that are brittle to surprise, fail to build consensus, undervalue long-term thinking, and undervalue the flexible and integrated management plans that are most effective at managing future risk. The paper presents a case study on how the RDM has been applied to ensure the robust flood management in HCMC. The RDM, a new approach to managing conditions of uncertainty, addresses the challenge of integrated flood risk management in HCMC. RDM has been increasingly used in developed countries, but this study provides the first application in a developing country. RDM seems very promising for HCMC: the city's challenges loom large in the face of tremendous uncertainty about the future and its planners have a strong interest in applying state-of-the-art planning and risk management methods to meet these challenges. In particular, the city's planners seek to implement a sophisticated integrated flood risk management strategy that combines innovative infrastructure, adaptation, and retreat policy options. In short, this paper provided insightful sharing experience to link strategic S&T interventions through RDM to mitigate flood related hazards due to extreme natural events in HCM City.

Keywords: *Robust decision making, Flood risk management.*

1. Introduction

Ho Chi Minh City (HCMC) ranks fourth globally among coastal cities most vulnerable to climate change (Nicholls at al. 2007). HCMC already experiences extensive routine flooding and in the coming decades increased precipitation and rising sea levels could permanently inundate a large portion of the city's population, place the poor in particular at risk, and threaten new economic development in low-lying areas (ADB 2010).

In response to this challenge, HCMC has embarked on a multi-billion dollar sewage and drainage infrastructure construction campaign. Over the last twenty years, the city has developed plans calling for the construction of 6000 km of canals and pipes – of which 1500 km have been built – and 172 km of dikes and river barriers.

Recently, the HCMC Steering Center of Urban Flood Control (SCFC) is revising the city's flood protection plans, in large part because increases in precipitation and tide levels observed over the last decade already exceed those projected. These unanticipated changes raise concerns that the original plans may not successfully manage flooding in HCMC and may even make it worsen in some areas. SCFC's new planning efforts aim to integrate and build upon two earlier plans (A 2001 Master Plan proposed 6000 km of canals and pipes. In 2009, the government also approved plans for roughly 172 km dikes and river barriers mainly for tidal control. The project to integrate these efforts, titled "***Technical Assistance Project for Ho Chi Minh City Flood and Inundation Management Project***," should run from 2009 to 2012. The Dutch Government has provided VND 42 billion to fund this study, equal to roughly 2 million US dollars) for drainage and flood control. The agency is also developing new, more sophisticated models to assess the city's flood risk more comprehensively and to support consideration of a broader array of policies and plans to manage these risks.

The pilot project supported by the World Bank could help SCFC incorporate a state-of-the-art approach ***Robust Decision Making*** **(RDM)** for managing uncertainty in this new planning effort. RDM will help ensure that HCMC's new flood control plans can successfully handle the wide range of climate and other changes that the city is sure to face in the decades ahead. The next section of the paper describes the RDM method and then explains how RDM might be incorporated into SCFC's planning process.

RDM improves planning under conditions of deep uncertainty and helps policy makers identify strategies that are robust over many plausible futures, rather than strategies that are optimal in any one best-estimate of the future but brittle in others. RDM most often proves useful when decision makers face conditions of deep uncertainty, where fast-changing and hard-to-predict future conditions leave decision makers unsure about the accuracy of their models or the likelihood of different future conditions.

To identify robust strategies, RDM conducts the analysis in reverse order compared to traditional risk analysis. Rather than begin with best-estimate projections of the future, RDM begins with one or more strategies under consideration. The analysis then examines the performance of the strategy under many future projections – hundreds or thousands — and identifies scenarios in which the proposed strategy meets its goals, and scenarios in which it does not. This scenario analysis reveals the robustness of the strategy. Using this information, RDM then helps decision makers identify new or augmented strategies that address those vulnerable scenarios, and evaluate whether those new strategies are worth adopting. RDM is not a new type of risk model. Rather, it provides innovative methods and tools that help planners use their models differently, and RDM itself is embedded in an explicit process of stakeholder engagement. Thus, RDM can be used with existing models and data, and can augment existing decision support processes.

In this case study RDM provides novel results using an existing flood risk model for Ho Chi Minh City (HCMC). The analysis helped facilitate and drew on the outcomes of two stakeholder workshops in HCMC with government authorities, local experts, consultants undertaking related studies, and representatives from international institutions.

2. Material and Methods

This paper demonstrates RDM analysis of food risk management in the NhieuLoc- ThiNghe (NL-TN) Canal catchment area in Ho Chi Minh City, shown in **Figure 7.1**. This area faces high flood risk and has received significant investment in flood risk management.

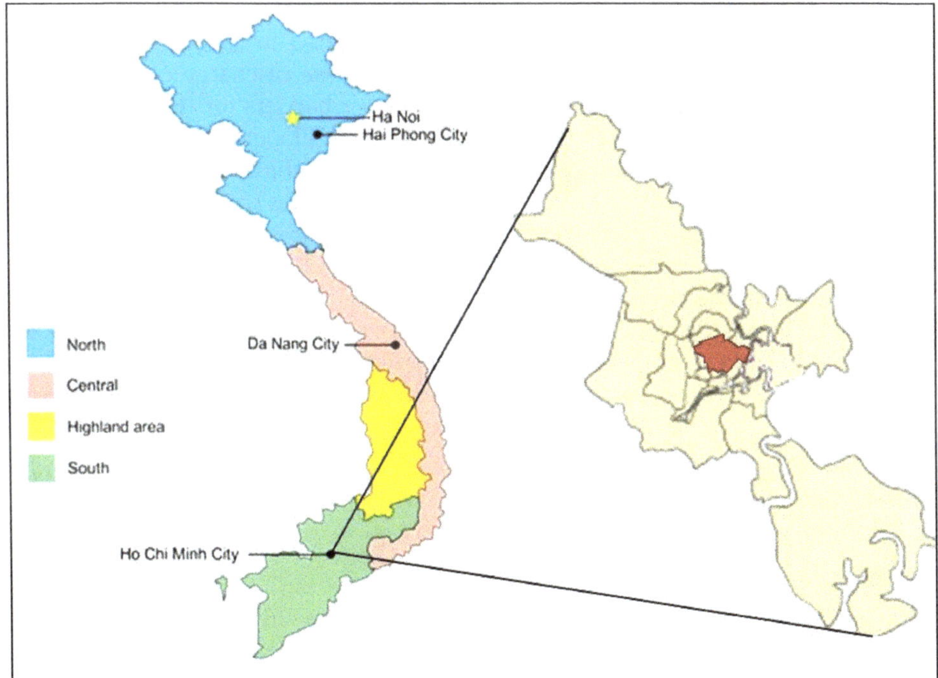

Figure 7.1: Vietnam with Ho Chi Minh City in South Vietnam Highlighted (Green) and Ho Chi Minh City with the NhieuLoc-ThiNghe Catchment Highlighted (Extended View).

The original infrastructure planning for NL-TN catchment did not include a full uncertainty analysis. As mentioned, in the years since, actual climate and socioeconomic conditions have changed significantly from what was projected at the time. Building on the Steering Centre for Flood Control's existing models and data, this study repeats the previous analysis, this time using an RDM framework to help manage uncertainty and help develop a robust, integrated plan. This choice of study design was motivated by two considerations. Firstly, this study aims to demonstrate how the Steering Centre for Flood Control and a wide range of other organizations can use RDM to augment their existing planning activities to improve their ability to manage uncertainty. Thus, it was important to build this study on models and data that the Steering Centre for Flood Control had previously used. Secondly, the Steering Centre for Flood Control is in the process of developing a more comprehensive integrated flood risk management strategy, using a new flood risk modeling system developed by the international engineering and project management consultancy ***Royal Haskoning*** and ***Deltares***- Netherlands Centre for Coastal Research. The plan resulting from this process will address flood risk over the entire city and is likely to be considerably more robust to future uncertainty than the infrastructure investments considered here. Nonetheless, the process used to develop this new plan does not yet take account of the full range of plausible climatic and socio-economic futures facing Ho Chi Minh City.

3. Results, Findings and Discussion

Over the last twenty years, HCMC has developed plans for and begun to implement numerous infrastructure projects designed to reduce its flood risk. In particular, 6000 km of canals and pipes to increase discharge capacity of the storm water system and 172 km of dikes and river barriers for tidal control will soon be completed. The city's Steering Committee on Flood Control (SCFC) seeks an innovative flood risk management strategy that combines these infrastructure investments with adaptation policies such as building codes and more porous urban surfaces, and retreat policies, such as concentrating housing and businesses on higher ground, while using lower lying lands for interruptible uses like recreation. However, evaluating such an integrated strategy, in particular its ability to manage uncertainty in a wide range of plausible futures proves difficult with traditional risk management methods. This analysis suggests that an integrated strategy to reduce flood risk in HCMC is needed and should be mainstreamed into the City urban planning.

RDM analysis and workshops have been conducted with a relatively simple ***Storm Water Management Model*** **(SWMM)** originally developed by SCFC and tailored to the NL-TN catchment.

Below are the findings:

- ☆ HCMC's soon-to-be completed infrastructure investments will reduce economic and population risk in the NL-TN catchment compared to recent levels in current and best estimate future conditions.
- ☆ This infrastructure may fail to reduce such risk in many future combinations of plausible socio-economic and climate trends.

- Augmenting the new infrastructure with adaptation and retreat measures can significantly expand the range of conditions for which risk is reduced.
- A risk management strategy that is explicitly designed to evolve over time can achieve almost the same level of risk reduction as a strategy that implements all such options immediately. In such an *adaptive* strategy, some policy options are implemented immediately and others implemented in the future only if specific warning signs suggest they become necessary.
- The adaptation and retreat options that reduce economic risk also reduce population risk, but significant additional measures are needed to reduce population risk by comparable levels.

3.1 RDM Improves Planning under Conditions of Deep Uncertainty

RDM helps policy makers identify strategies that are robust over many plausible futures, rather than strategies that are optimal in any one best-estimate of the future but brittle in others (***As described in the IPCC Special Report on Managing the Risks of Extreme Events and Disasters to Advance Climate Change Adaptation, SREX report, resilience is a property of a system while robustness is a property of a decision***). RDM most often proves useful when decision makers face conditions of deep uncertainty, where fast-changing and hard-to-predict future conditions leave decision makers unsure about the accuracy of their models or the likelihood of different future conditions. To identify robust strategies, RDM conducts the analysis in reverse order compared to traditional risk analysis. Rather than begin with best-estimate projections of the future, RDM begins with one or more strategies under consideration. The analysis then examines the performance of the strategy under many future projections hundreds or thousands and identifies scenarios in which the proposed strategy meets its goals, and scenarios in which it does not. This scenario analysis reveals the robustness of the strategy. Using this information, RDM then helps decision makers identify new or augmented strategies that address those vulnerable scenarios, and evaluate whether those new strategies are worth adopting.

3.2 RDM Provides Important Value-Addition

These findings are derived from very simple simulation models and an extremely simplified representation of the impacts of the alternative policies. Thus, these findings are almost certainly subject to revision if and when revisited with a more detailed model. This analysis nonetheless helps demonstrate the benefits RDM brings to the challenge of developing effective risk management strategies in fast-changing and deeply uncertain developing country environments.

In this study, RDM provides two sets of tangible products:

1. ***Scenarios*** that illuminate the vulnerabilities of proposed policies. These scenarios emerge from the analysis and provide a detailed understanding of the combinations of future climate and socio-economic conditions where a proposed flood risks management strategy will and will not meet its risk reduction goals. These scenarios help planners compare the robustness

of competing strategies and identify augmentation options that could address vulnerabilities.

2. ***Trade-off curves*** that concisely summarize the relationship between cost and risk reduction provided by each of several flood risk management strategies. These curves help planners consider the impacts of different strategies on different populations and between economic and population risk.

These products and the process that produces them help SCFC and other decision makers in HCMC to:

1. Implement a full and systematic identification of risk and compare response options even when significant shortcomings in the available climate and socioeconomic data exist.
2. Consider adaptive strategies, which may reduce deeply uncertain risk more effectively by evolving over time in response to new information.
3. Facilitate structured discussions among stakeholders regarding vulnerability and tradeoffs even in the face of deep uncertainty

With these attributes, RDM analyses can help decision makers develop strategies that will prove successful over a wide range of unexpected and potentially surprising futures, and help facilitate the extensive stakeholder interactions needed to build consensus for such strategies.

3.3 Promising Conditions and Pathways Exist for Adopting RDM

In general, RDM proves most useful when (as in HCMC), decision makers:

1. Face conditions of deep uncertainty, such as fast-changing and hard-to-predict climate and socio-economic conditions.
2. Have a diverse set of policy options that, in various combinations, could contribute to policies that are robust to those uncertainties.
3. Have or can develop simulation models that relate policy options to their potential consequences and
4. Are motivated to adopt innovative planning approaches.

To successfully adopt RDM, organizations must be willing to augment their planning processes to incorporate new types of quantitative information *e.g.* scenarios that illuminate vulnerabilities — to manage risk. RDM also asks decision makers to shift from predict-then-act analyses, in which planners begin with best-estimate projections of the future, to vulnerability-and-response analyses, in which planners assess the vulnerabilities of one or more strategies under consideration.

Integrating RDM into an organization's planning efforts can yield important benefits, including allowing them to:

1. Fully and systematically identify risks and compare risk management options, even when there exist significant shortcomings in the available climate and socioeconomic data

2. Consider adaptive strategies, which may reduce deeply uncertain risk more effectively by evolving over time in response to new information and
3. Facilitate structured discussions among stakeholders regarding vulnerability and tradeoffs even in the face of deep uncertainty.

HCMC is one of many places using RDM to improve its decision support and long term planning. The State of California uses RDM to facilitate the development of its long-range water management plan, the U.S. government is using the approach to develop management strategies for the Colorado River, and RDM contributed to the State of Louisiana's and City of New Orleans' recent master plan for a sustainable coast. The experience with RDM's use in the United States suggests how organizations in HCMC and other developing countries might most effectively adopt RDM.

The diffusion of RDM concepts and methods often begins with a demonstration project for an organization in a particular sector that regards itself as a thought leader and first adopter of new methods. Often this organization has a particularly difficult planning challenge that other methods have been unable to solve satisfactorily. The organization first uses RDM in a planning activity related to, but not on the critical path of its formal planning activities. In some cases, the organization itself funds the pilot study and in others initial support comes from an outside institution. After conducting the pilot study, the organization will begin to fold RDM into its ongoing planning activities, while developing its own internal capability to implement RDM analyses. Organizations following this pattern include the California Department of Water Resources, the Metropolitan Water District of Southern California, and the U.S. Bureau of Reclamation.

This HCMC demonstration effort follows the initial steps of this general pattern. HCMC faces significant flood risk management challenges, the city regards itself as a thought-leader in addressing these challenges, and it has engaged in this RDM exercise with support from the World Bank and Korean Global Green Growth Initiative.

4. Conclusion

Ho Chi Minh City faces significant future flood risk driven by hard-to-predict future climate and socio-economic trends. Many regions in the world face similar risk management challenges. However, traditional quantitative methods for risk management often prove inadequate because they can produce strategies that are brittle to surprise, fail to build consensus, undervalue long-term thinking, and undervalue the flexible and integrated management plans that may be most effective at managing future risk. The paper presents the case study on how the RDM has been applied to ensure the robust flood management in HCMC. The RDM is a new approach to managing conditions of uncertainty and addresses the challenge of integrated flood risk management in HCMC. RDM has been increasingly used in developed countries, but this study provides the first application in a developing country. RDM seems very promising for HCMC: the city's challenges loom large in the face of tremendous uncertainty about the future, and its planners have a strong interest in applying state-of-the-art planning and risk management methods to meet these challenges. In particular, the city's planners seek to implement a sophisticated integrated flood risk management strategy that combines innovative infrastructure, adaptation, and retreat policy options. In short, this paper provides insightful sharing experience to link strategic S&T interventions through RDM to mitigate flood related hazards due to extreme natural events in HCM City. The methods demonstrated in HCMC could also prove broadly applicable in many regions of the world. In its 2010 World Development Report and other publications, the World Bank recommends RDM as a widely useful approach to managing risk and uncertainty associated with development and future climate change.

Acknowledgements

The paper is written based on the background and publications from the Project *"**Developing Robust Flood Defenses for Ho Chi Minh City**"* supported by the World Bank.

References

1. ADB 2010. Ho Chi Minh City Adaptation to Climate Change. Mandaluyong City, Philippines, Asian Development Bank, 2010. https://www.adb.org/sites/default/files/publication/27505/hcmc-climate-change-summary.pdf
2. Lempert, R. Kalra N., Peryraud S., Zhimin M., Sinh B., Dean C. and Alexander L. 2013. Ensuring robust flood risk- management in Ho Chi Minh City. Policy Research Working Paper from the World Bank # 6465, Washington DC, World Bank.

   https://www.researchgate.net/publication/255698092_Ensuring_Robust_Flood_Risk_Management_in_Ho_Chi_Minh_City
3. Nicholls R.J., Hanson S., Herweijer C., Patmore N., Hallegatte, S., Corfee-Morlot J., Jean C., and Muir-Wood R. 2007. Ranking of the World's Cities Most Exposed to Coastal Flooding Today and in the Future. Paris, France: OECD publishing.

   https://climate-adapt.eea.europa.eu/metadata/publications/ranking-of-the-worlds-cities-to-coastal-flooding/11240357

Chapter 8

An Assessment of Land Use Dynamics in Pallikaranai Wetlands, Tamil Nadu

K.S. Vignesh[1], V.M. Suresh[2], K. Leelavathy [1] and S.V. Varshini[3]*

[1]University of Madras, Chennai – 600 005, Tamil Nadu, India
[2] CNHDS, University of Madras, Chennai – 600 005, Tamil Nadu, India
[4]Department of Civil Engineering, Jerusalem Engineering College, Chennai – 600 100, Tamii Nadu India
**E-mail: ksvigneshphd@gmail.com*

ABSTRACT

Wetlands which can be the ecosystems or habitat for unambiguous animals and plants are saturated with water, so it is clear that the existence of water predetermines their formation, processes and characteristics. Wetlands considered as a great importance to nature and human being which has a tendency to purify the water, flood reduction, storage of water, and also mitigate the soil erosion and attains support to varied biodiversities. Pallikaranai Wetlands Ecosystem is one of the known wetlands of the world which has been facing the menace of urban expansion and population pressure. This wetlands being present in the periphery of Chennai city which is rapidly expanding makes it more vulnerable to human misuse. This paper aims to study the impact on the wetlands and land use dynamics, most modern scientific method has been adopted, using sophisticated software ARCGIS, ERDAS. Land use analysis for the past thirty years shows that the current wetlands have diminished in area. The most important step will be to stop using the wetlands as dumping site and realize its utility and change the perception of wetlands and its rich biodiversity from a place to be used as a dumpsite to a place which can save humanity.

Keywords: *Wetlands, Biodiversity, Urban expansion, Land use dynamics, Wasteland, Sewage treatment plant.*

1. Introduction

The assessment of millennium ecosystem, a conceptual framework for ecosystems and well human-being has acknowledged the importance of wetlands in the context of providing a framework that supports the promotion and delivery of the Ramsar Convention's "wise use" concept. This existing guidance provided by the convention for the wise use of all wetlands to be expressed within the context of human wellbeing and poverty alleviation. The adverse effects of climate change, such as sea level rise, coral bleaching, and changes in hydrology and in the temperature of water bodies, will lead to appropriate reduction in the services provided by wetlands. The most effective method of coping with the adverse effects of climate change is by removing the existing pressures on wetlands and improving their resiliency. Conserving, maintaining, or rehabilitating wetland ecosystems can be a viable element to an overall climate change mitigation strategy (MEA, 2005). Wetlands are ecosystems or habitats for specific plants and animals that are saturated with water, the presence or absence of water determines their formation, processes and characteristics (Shaw and Fredine 1956, Cowardin *et al.*, 1979, Zoltai 1988, Finlayson and Moser 1991). Wetlands are of great importance to man and nature as it purifies water, reduces flood, stores water, prevents soil erosion and supports varied biodiversities (Smith *et al.*, 1994, Massel *et al.*, 1999, Katharesan and Rajendran 2005). Starting about thousands of years in urban areas of the world and typically a few hundred years ago in most of the coastal areas, humanity has profoundly impacted, degraded or destroyed many coastal wetlands worldwide by direct physical degradation and pollution (Streever 2001, Zong *et al.*, 2007, Wolanski 2007). Ironically, reduced coastal wetland increases threat to human safety at the same time that shoreline development exposes populations to coastal hazards such as tsunamis, erosion, flooding, storm and waves surges (Katharesan and Rajendran 2005). Added to this, these wetlands are store house of vivid biodiversity (Neill 1958). Numerous waterfowl live in the wetlands (Goss-Custard *et al.*, 1977, Erwin 1996). Wetlands are habit of some rare species which play a vital role in the ecosystem and the food chain (Mitsch and Gosselink, 2007). Destruction of wetlands will destroy the habitat of numerous species of flora and fauna that live in the Wetlands.

A land-use and land-cover play a major role in the changes of any natural environment. Land use/land cover changes by human and through natural means; have resulted in deforestation, biodiversity loss, global warming and increase natural disaster mainly flooding (Rogana and Chen 2004). The present study explores the Pallikaranai Wetland of Tamil Nadu State in India for the assessment of land-use and land-cover dynamics to develop a predictive framework for the change in local habitat and associated environment.

2. Study Area

Chennai Metropolitan Area (CMA) is the fourth largest in India and 30th in the world in terms of population. The CMA consists of the metropolitan city of Chennai and its suburbs spread in Thiruvallur and Kancheepuram districts of Tamil Nadu. It lies between 12° 50′ N to 13° 17′ N longitude and 79° 59′E to 80°

19′ E latitude. Pallikaranai is a part of CMA and is located just south of Chennai City in Kancheepuram District. After the expansion of the Chennai Corporation, Pallikaranai will fall under zone fourteen of the Chennai Corporation (Corporation of Chennai 2013), and houses the famous Pallikaranai wetlands. The Pallikaranai wetland is presently being polluted due to indiscriminate dumping of Municipal Solid Waste (MSW) and disposal of sewage waste directly in to the wetlands.

3. Materials and Methods

The remotely sensed satellite images are processed for the identification of land-use changes around the study area. The ***Level-I*** classification has been done for the assessment of land-use dynamics. They are urban or built-up area, agricultural land, rangeland, forestland, water, wetland, barren land, tundra, perennial snow or ice.

4. Result and Discussion

4.1 Land-Use of Pallikaranai-1991

The land use data of the year 1991 shows large area under settlement which accounts for around 39 per cent of the total area, it is followed by wetlands which occupies 26 per cent of the area. Water bodies cover 3.6 per cent of the area, where as the vegetation covers 17.4 per cent of the area and waste land and crop land covers 11, and 2.2 per cent of the area respectively. Sewage Treatment Plant (STP) was not established by this period of time. The area for the land use is clearly illustrated in the **Figure 8.1**.

4.2. Land-Use of Pallikaranai-2001

The land-use data of the year 2001 shows large area under settlement which accounts for around 39 per cent of the total area, it is followed by wetlands which occupies 26 per cent of the area. Water bodies cover 5.3 per cent of the area, where as vegetation covers 15.4 per cent of the area and waste land covers 1.5 per cent of the area while, cropland covers 11.5 per cent of the area. In 2001 the area under cropland has increased, while a decrease in the area under waste land was also observed. The area for the land use is clearly illustrated in the **Figure 8.2.**

4.3. Land-Use of Pallikaranai-2006

The land-use data of the year 2006 shows large area under settlement which accounts for around 57.4 per cent of the total area it is followed by wetlands which occupy 18 per cent of the area. Water bodies cover 5 per cent of the area, where as vegetation covers 6 per cent of the area and waste land covers 8.9 per cent of the area. Cropland covers 2.9 per cent of the area. In 2001 the area under cropland has decreased, large area under wetland has been reduced and area under settlement and wasteland has increased. The area for the land-use is illustrated in the **Figure 8.3.**

4.4. Land-Use of Pallikaranai-2010

The Land-use data of the year 2010 shows large area under settlement which accounts for around 63.33 per cent of the total area it is followed by wetlands which occupy 13.14 per cent of the area. Water bodies cover 3.16 per cent of the area,

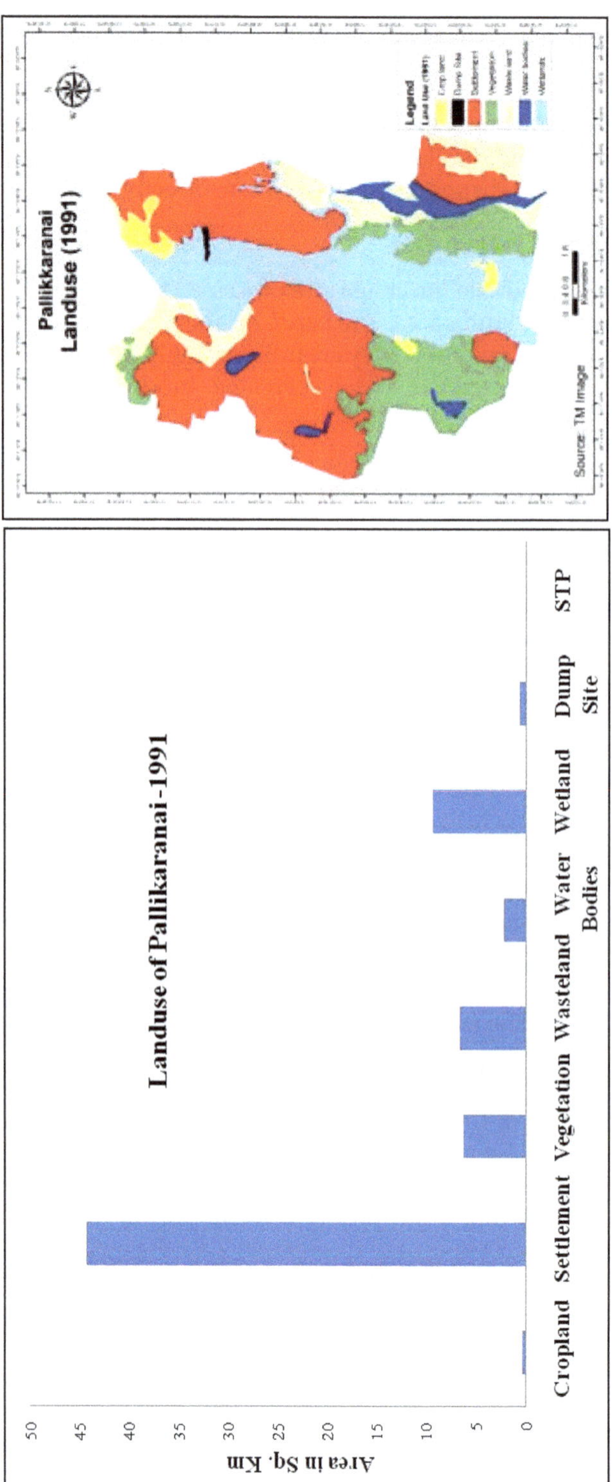

Figure 8.1: Land Use of Pallikarani in Year 1991.

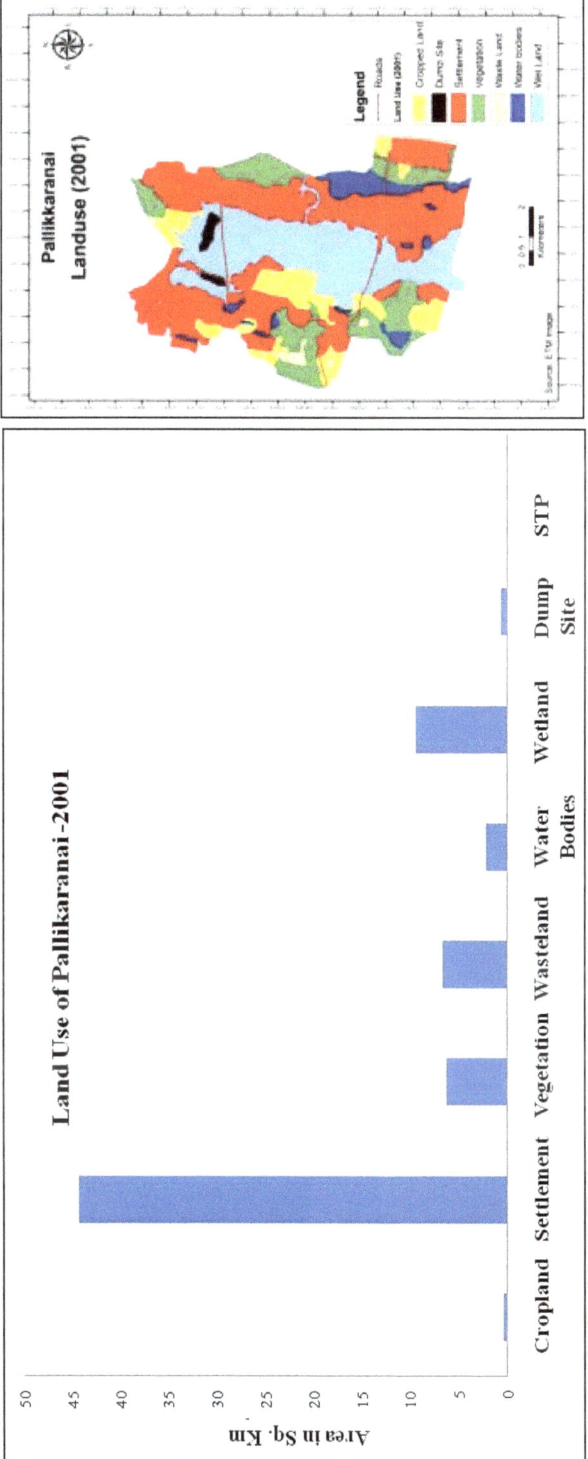

Figure 8.2: Land Use of Pallikarani in Year 2001.

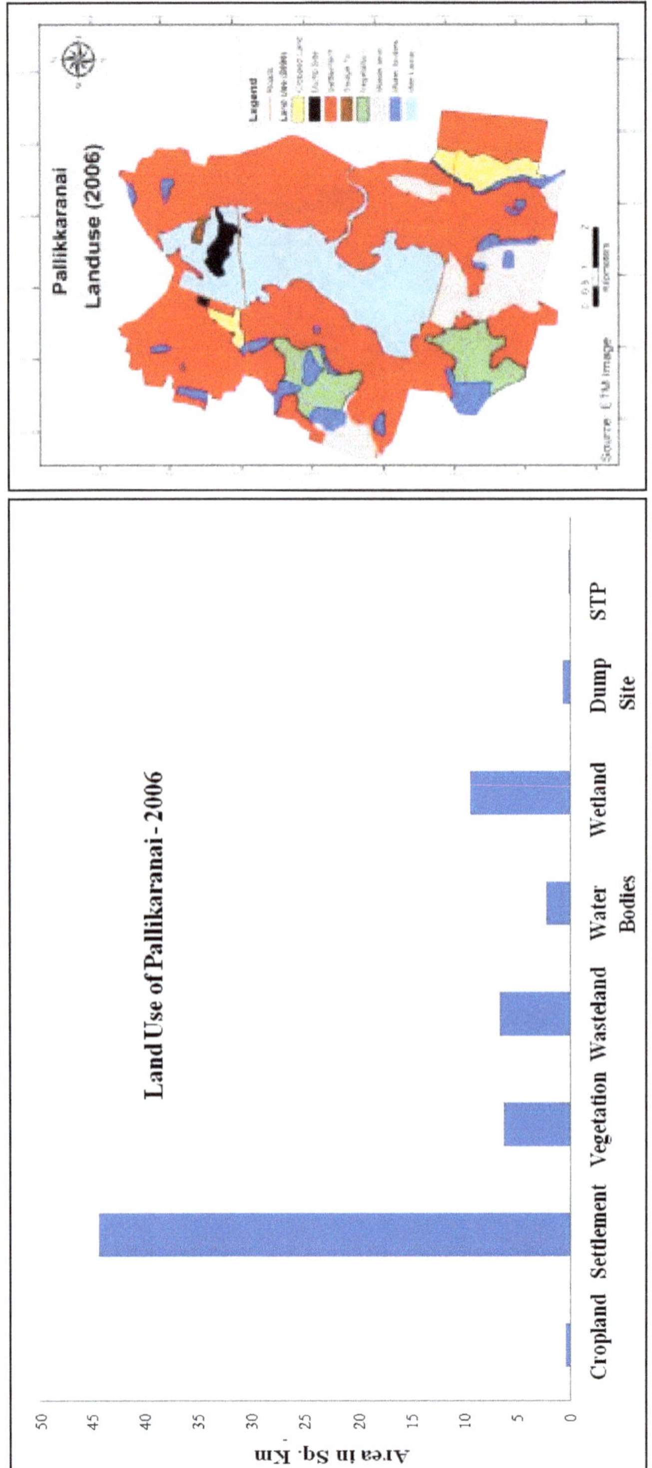

Figure 8.3: Land Use of Pallikarani in Year 2006.

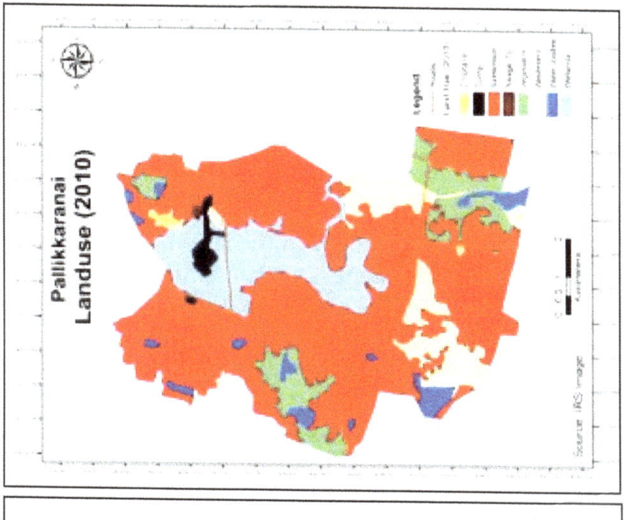

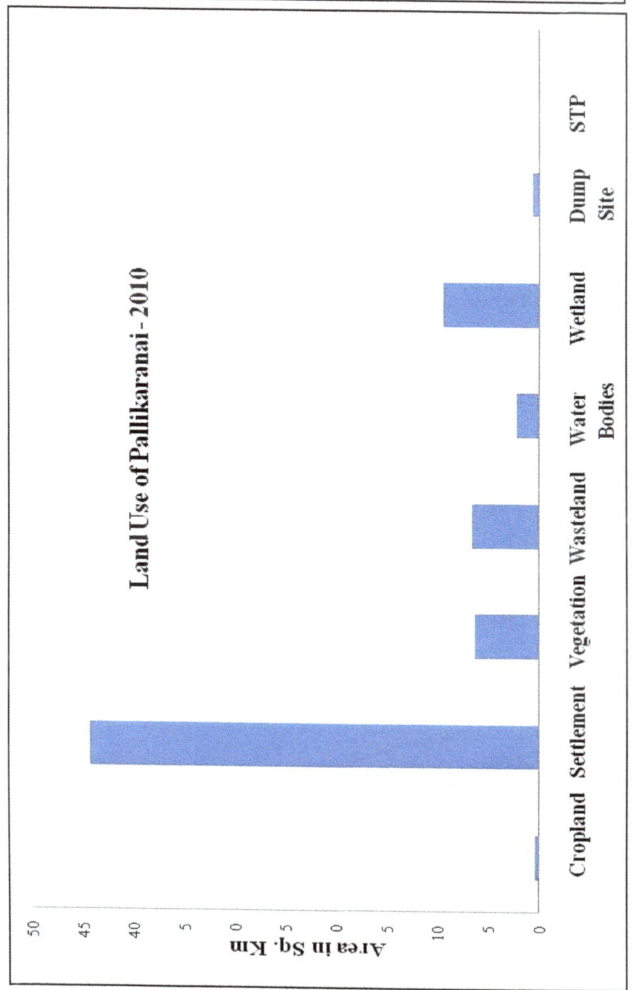

Figure 8.4: Land Use of Pallikarani in Year-2010.

where as vegetation covers 8.96 per cent of the area and waste land covers 9.45 per cent of the area. Cropland covers just 0.54 per cent of the area. It can be seen that the area under wetlands have been decreasing steadily whereas the area under settlement and dump site has been constantly increasing. The area for the land-use is illustrated in the **Figure 8.4**.

4.5. Land-Use Dynamics

The **Figure 8.5** illustrates the changes in the land use from 1991 to 2010. It can be seen that the area under settlement is on a continuous rise whereas the area under wetlands is on the decline. The area under crop land showed an increasing trend in the year 2001, which may be due to truck farming, but in the subsequent years, it also declined rapidly to give way to settlement demands. The area under vegetation has increased marginally in the year 2010 from a consistent decline from 1991. The area used for dumping increased in 2001 and it has seen unchanged in horizontal expand, but a field visit indicates vertical increase in the dump area, and also due to the growth of vegetation in the dump yard, some areas in the dump yard are covered by vegetation. All other Land use are on the decline, but the area under waste land has a fluctuating trend and is on the increase, and it may be due to speculation and real estate business which is on sharp rise due to rapid industrialization of the area.

4.6. Change Detection

The change in the land-use compared with 1991 to that with 2010 shows marked reduction in area of wetland and vegetation. Wetlands have reduced by 12.8 per cent whereas vegetation has reduced by 8.5 per cent. Cropland, wasteland, area under water bodies have also showed a marked reduction. The reduction in area

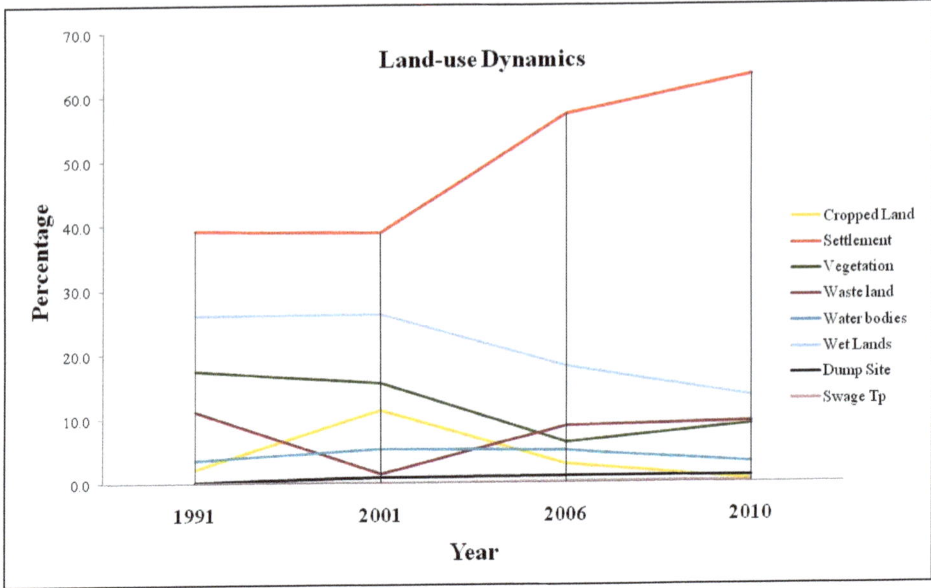

Figure 8.5: Land Use Dynamics Observed from 1991-2010.

of other land-use mentioned above is occupied by settlement which has increased by 24 per cent in the subsequent decade. The area under dump site and sewage treatment plant (STP) has also shown a marked increase **Figure 8.6**.

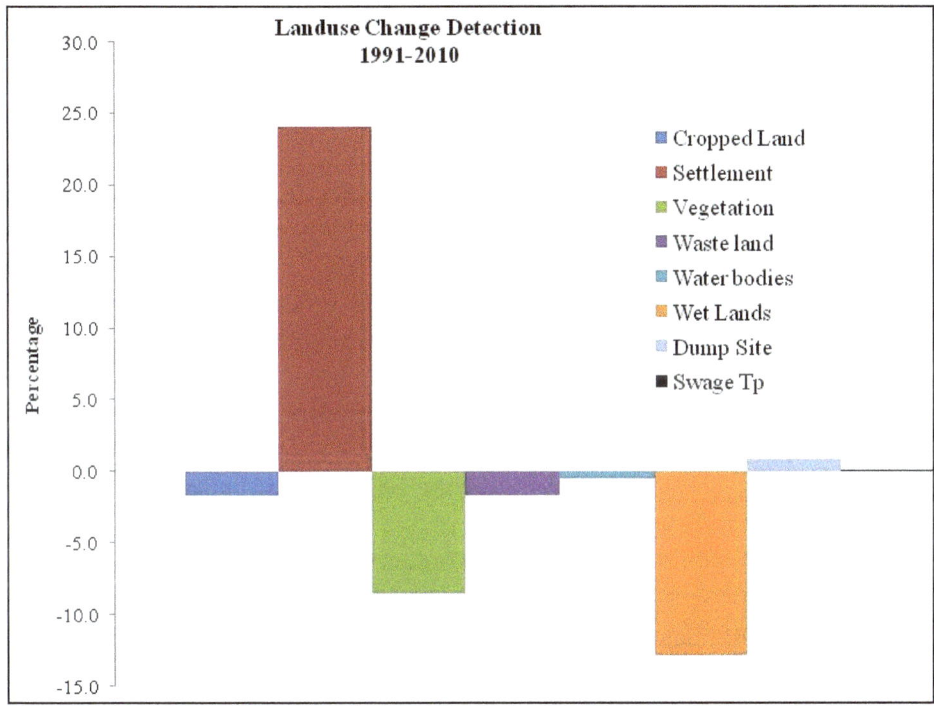

Figure 8.6: Land Use Change Detection Observed for 1991-2010 Decade.

5. Conclusion

The land-use changes are one of the most important factors in determining the impact on the environment and its biodiversity. To understand the impact of different factors relating to the land-use dynamics on the Pallikaranai wetland and its biodiversity, the pattern of land-use was calculated using Satellite Imageries and Image processing software. GIS software was used to determine and map the land use changes. From the data obtained from satellite imageries, it can be seen that the area under settlement is on the rise and the area under wetlands is declining consistently. The area under solid waste disposal site is also on the increase in the initial period, though the rise in the area of the solid waste disposal site has been restricted due to steps taken by different organizations. The area under cropland has been changing and has reduced to a meager level, same is the fate of vegetation and water bodies which have seen a study decline, though the area under waste lands is fluctuating and this may be due to speculation in the real state marked where in large area of land are left fallow for the speculation of price and are later converted into settlement sights. The change detection graph clearly shows negative growth of all types of land use except that of settlement and dump site area, when comparing the land use of the year 1991 with that of year 2010.

Acknowledgment

We would like to acknowledge the staff members of Chennai District Collector Office, Greater Chennai Corporation and Tamil Nadu Pollution Control Board, for gladly sharing the data for the present work.

References

1. Pacifica O.F. 2007. Environmental impact assessment general procedures. Presented at Short Course II on Surface Exploration for Geothermal Resources, UNU-GTP and Ken Gen, Lake Naivasha, Kenya.
2. Adamus P.R. 1983. A method for wetland functional assessment. U.S. Department of Transportation, pp. 16-134, Washington, DC, USA
3. Adisak T. 1982. Solid waste management in regional cities of Thailand. Paper presented at Regional Seminar on Solid Waste Management, Thailand.
4. Adomokai Rosemary and William R. S. 2004. Community participation and environmental decision-making in the Niger Delta. *Environmental Impact Assessment*. 24: 495-518.
5. Ahmad Y. J. and Sammy G. K. 1987. Guidelines to environmental impact assessment in developing countries. *UNEP Regional Seas Reports and Studies*.
6. Amezaga J.M., Santamaria L., Green A.J. 2002. Biotic wetland connectivity supporting a new approach for wetland policy. *Acta Oecologica* 23: 213-222.
7. Anjaneyulu Y. and ValliManickam V. 2010. Environmental impact assessment methodologies, BS Publications, Hyderabad
8. Black R.J and Weaver L. 1967. Action on the solid waste problem. *Journal of the Sanitary Engineering Division* 93: 437-446.
9. Brooks R.P., Samuel D.E. and Hill J.B. eds. 1985. Wetlands and Water Management on Mined Lands. Proceedings of Conference, University Park: Pennsylvania State University Press, pp. 393., USA
10. Frayer W.E., Monahan T.J., Bowden D.C., Graybill F.A. 1983. Status and trends of wetlands and deep water habitats in the conterminous United States. 1950's to 1970's, Colorado State University, Ft. Collins, CO, USA.
11. Lu J. 1995. Ecological significance and classification of Chinese wetlands. *Vegetation* 118: 49-56.
12. Salem H.M., Eweida A.E. and Farag A. 2000. Heavy metals in drinking water and their environmental impact on human health. Cairo University, ICEHM, pp.542- 556. Egypt
13. Shutes R.B.E, Revitt D.M., Mungar A.S. and Scholes L.N.L. 1997. The design of wetlands systems for the treatment of urban runoff. *Water Science and Technology* 35:19-25

14. Sunilson J.A.J, Rejitha G., Anandarajagopal K. 2009. Cytotoxic effect of *Cayratiacarnosa* leaves on human breast cancer cell lines. *International Journal Cancer Research* 5: 1-7

15. Tucker C.S. and Louis R.D. 2008. Managing High pH in Freshwater Ponds. *Southern Regional Aquaculture Centre* (*SRAC*) *Publication No. 4604,* USA.

Chapter 9

The Use of Remote Sensing and GIS for Drought Risk Assessment: The Case of Southern Province, Zambia

Micheal Katongo Phiri[1,2]

[1]*National Remote Sensing Centre, Chelstone, Lusaka, Zambia*
[2] *IWRM Centre/Department of Geology, School of Mines, University of Zambia, Great East Road Campus, P.O. Box 32379, Lusaka 10101, Zambia*
E-mail: mikaphi@yahoo.com

ABSTRACT

Water plays a major role in human livelihood and has significant implications for security. Hydrological extremes, such as flood and droughts are the major threats to society and can have extensive effects. Water resources are already under tremendous pressure due to population growth, economic activities and escalating competition between end users, and droughts present several challenges to water resource managers. Remote Sensing and Geographic Information Systems (GIS) have emerged as essential tools in the assessment and analysis of natural resources and hazards (natural and man-made). In this study, multi-temporal images from the MODIS Terra (250m) were used to derive the Normalized Difference Vegetation Index (NDVI) of the Southern Province of Zambia (2000 to 2017). Rainfall data from five meteorological stations were obtained and used in the correlation analysis with the NDVI. Correlation analyses were done between satellite-derived NDVI and rainfall. A positive correlation (r = 0.55) was found between NDVI and rainfall. NDVI was found to be fairly correlated with the mean annual rainfall of the Southern Province. Rainfall variation displayed a north to south gradient, with variability increasing towards the south. On the other hand, agricultural drought exposure followed a pattern more similar to land cover. The findings of this component will be an input for further assessments of the drought risk.

Keywords: *NDVI-rainfall correlation, Rainfall variability, Drought.*

1. Introduction

1.1 Background

Water is essential for life and has significant implications for security. Hydrological extremes, such as droughts and floods, are major threat to the society and can have extensive effects (Kundzewicz and Matczak, 2015). Zambia, which is not immune to drought issues, is largely dependent on rain fed agriculture with the largest part of the grain crops produced by small scale farmers. However, insufficient public investment in small scale agriculture has left the country in a very difficult position during times of drought and water scarcity (Lekprichakul, 2008). The Southern Province, being one of the largest producers of maize and other crops in the country, does not receive as much rainfall as the provinces in the north and thus it is prone to drought. It is, therefore, important to assess drought risk in the area using advanced methods such as Remote Sensing and Geographic Information Systems (GIS).

1.2 Definition of Drought

In 1965, the Director of Commonwealth Bureau of Meteorology suggested a general definition of drought as "Severe Water Shortage".

1.3 Statement of the Problem

In Zambia, over the past two decades, the arid conditions within the Southern Province have increased the drought risk in the area. With the increasing pressure on water resources from a growing population, economic activities and ever-increasing competition between users, it is important to evaluate the possible risk arising from drought in order to aid in developing better water management plans and drought impact mitigation measures.

Conventional drought risk assessment techniques that are currently used require a well populated network of meteorological stations. However, the Southern Province only has 11 operational stations that are sparsely distributed across the province. To populate a network of meteorological stations for a large region such as the Southern Province requires large scale financing. Furthermore, collecting reliable data from the stations can prove cumbersome and would require a large number of trained individuals to accomplish (in case of manual meteorological stations). Therefore, it is important to consider other alternatives that can supplement or substitute conventional drought risk assessment techniques.

Remote sensing and GIS drought assessment techniques are not reliant on station based meteorological data. Satellite information is readily available and can be collected at high resolutions and almost in real-time from many free satellite platforms. Therefore, remote sensing and GIS can be used as a cheap alternative or supplement to conventional drought assessment techniques. Hence, it is important to explore the use of remote sensing and GIS to assess drought risk in the Southern Province.

2. Methodology

2.1 Description of Study Area

The Southern Province is located in the southern region of Zambia. It is located between the latitudes 15°S and 18°S and the longitudes 25°E and 29°E (**Figure 9.1**). There are 13 districts in the Southern Province and the provincial capital is Choma. Livingstone is the only city in the province and is the tourist capital of Zambia.

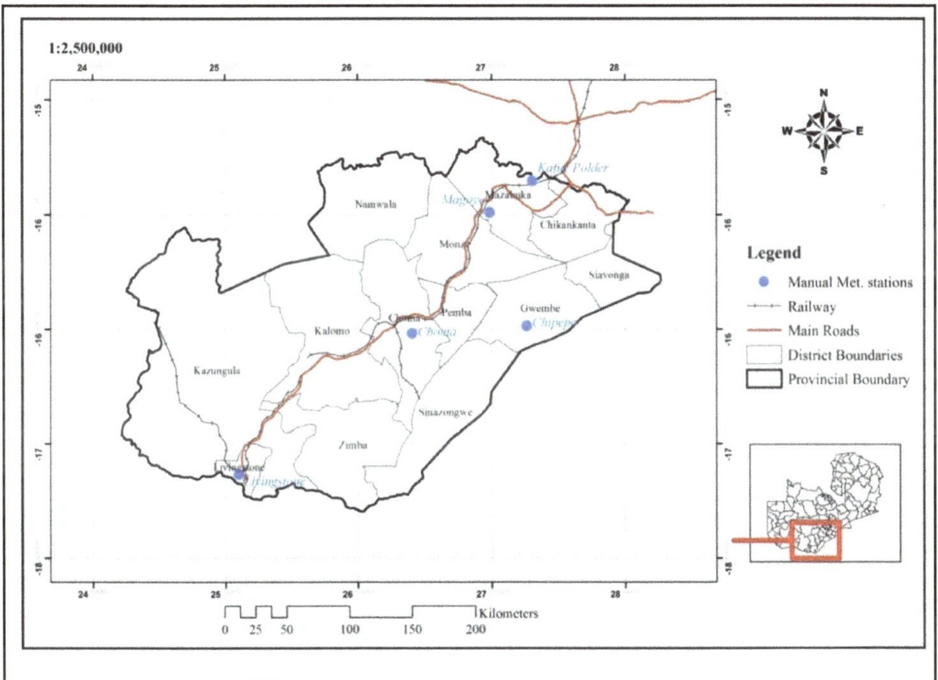

Figure 9.1: Southern Province, Zambia.

Zambia has four agro-ecological zones, two of which are found in the Southern Province. These are Region-I and IIa. Agro-ecological Region-I stretches laterally the western and southern parts of the province, along the Zambezi and the Luangwa Escarpment. The region sits on valleys and plains that are less than 1000 m above sea level. The region has limited production potential due to poor sandy soils (FEWS NET, 2014). Average annual rainfall is less than 800 mm and the growing season is short. Furthermore, temperatures are relatively higher than other places in the country. The region is most suitable for small scale farming of drought resistant crops such as millet, sorghum, sesame and cotton (FEWS NET, 2014).

Agro-ecological Region II is at the centre of the province. The region sits on a plateau and altitudes vary from 1000 m to 1400 m above sea level. Annual rainfall is 800 mm to 1000 mm and the growing season ranges from 120 to 160 days, which supports growth of medium to long-term crop varieties (FEWS NET, 2014). The region is divided into two, Region-IIa and IIb. The Southern Province is only covered by Region-IIa. Along with the central and eastern plateaus, the southern plateaus are considered to have the best agricultural potential in Zambia because the land is suitable for the growth of a diversified base of primarily rain fed crops and livestock production (FEWS NET, 2014). The land is laid with fertile soil that supports maize, wheat, soybean and tobacco.

Remote sensing and GIS have emerged as essential tools in the assessment and analysis of natural resources and hazards, both natural and man-made. In this study, multi-temporal images from the MODIS (Moderate Resolution Imaging Spectroradiometer) Terra Satellite (250 m resolution) were used to derive and evaluate the Normalized Difference Vegetation Index (NDVI) of the Southern Province of Zambia from 2000 to 2017. The obtained NDVI were used to derive the Vegetation Condition Index (VCI) which in turn used to identify the areas prone to drought and to establish a trend of drought occurrence over the study area. Rainfall data from five meteorological stations were used to determine whether a correlation with the NDVI existed.The rainfall data was also used to calculate Standardized Precipitation Index (SPI) (Mckee, *et al.*, 1993) which was used to determine drought event severity. TAMSAT (Tropical Application of Meteorology Using Satellite) decadal rainfall estimates were used to calculate Rainfall Anomaly (RFA) percentage and to determine rainfall variability across the Southern Province of Zambia.

2.2 TAMSAT Data Processing

TAMSAT is derived from a combination of Geostationary Meteosat Data and gauge observations using a calibration approach that exploits both data sets (Tarnavsky, *et al.*, 2014). The rainfall estimates are also compared with six other satellite based rainfall datasets. The primary objective has been to render historic and near real-time precipitation estimates for drought monitoring in Sub-Saharan Africa where there is a lack or complete absence of *in-situ* observations (Maidment, *et al.*, 2014).

Rainfall anomaly percentage was calculated for each rainy season for the years 2000 to 2016 using TAMSAT rainfall estimates. This was computed using the following equation:

$$RFA = (ppt_i - avg.ppt)/avg.ppt * 100 \qquad Eqn.1$$

Where RFA is rainfall anomaly percentage, ppt_i is the precipitation of a given season and avg.ppt is the average seasonal precipitation for the study period. RFA percentage was classified into classes of plus or minus 10, 25 and 50 per cent.

Average seasonal rainfall was calculated in order to assess spatial (areal) variability, which is how rainfall amounts vary at different locations across the region. Temporal variability of precipitation within the province was determined using coefficient of variation, which is computed by:

$$Coefficient\ of\ variation = \sigma/avg.ppt \qquad Eqn.2$$

Where, σ is the standard deviation of rainfall and *avg.ppt* is the mean rainfall.

2.3 MODIS Data Processing

The imagery was first geometrically corrected from sinusoidal to geographic projection. After that, each scene was inspected for cloud cover. Images that had over 70 per cent cloud cover were discarded. The scenes with less cloud cover were clipped to the extent of the Southern Province and used to calculate NDVI using *eqn.1*. NDVI was calculated manually using the band math module in ENVI 5.1 software (*Exelis Visual Information Solutions, Boulder, Colorado*). The following expression was used:

$$(b1 - b2)/(b1 + b2) \qquad \text{Eqn. 3}$$

Where *b1* was the NIR band and *b2* is the RED band.

Furthermore, the yearly average NDVI were derived and used to generate a Maximum Value Composite (MVC). MVC is a multi-temporal approach used to derive VCI, Temperature Condition Index (TCI), Vegetation Health Index (VHI) and Temperature Vegetation-Dryness Index (TVDI). In this study, the approach was used to derive VCI. The MVC was created using yearly average NDVIs from the year 2000 to 2016. MVC was generated by layer stacking yearly average NDVI in ENVI 5.1, which was then used to calculate the $NDVI_{min}$ and $NDVI_{max}$ in Band Math module of ENVI 5.1:

$b1<b2<b3<\ldots\ldots bn$ $\qquad$ $b1>b2>b3>\ldots.bn$

Where, *b1* are the pixel values of the first year and *bn* are the pixel values of the nth year in the time series. The obtained $NDVI_{min}$ and $NDVI_{max}$ were used to calculate VCI.

VCI is calculated using *eqn.2*. With the use of band math, the difference between $NDVI_{max}$ and $NDVI_{min}$ calculated by the following expression:

$$b3 = b1 - b2 \qquad \text{Eqn. 4}$$

Where, *b1* is $NDVI_{max}$ and *b2* is $NDVI_{min.}$ After that, the following expression was used to calculate VCI.

$$VCI = (b1 - b2)/b3 \qquad \text{Eqn. 5}$$

Where *b1* is $NDVI_{current}$, *b2* is $NDVI_{min}$ and b_3 is the difference between $NDVI_{max}$ and $NDVI_{min}$.

Drought exposure based on VCI was determined using R-ArcGIS packages, which is an integrated platform of two different statistical packages R and ArcGIS. R is a software suite for data manipulation, calculation and graphical display (Venables, *et al.*, 2018) while, ArcGIS comprises of several routines, such as ArcMap, ArcCatalogue and the ArcToolbox. VCI of drought years was used to determine agricultural drought exposure. A 1×1 km grid covering the Southern province was created using the fishnet tool in the ArcToolbox. A point feature was created at the center of each grid. Using the Extract Multiple Values Tool in the Spatial Analyst Toolbox, values of drought year VCI were extracted to each point. Coordinates of each point were generated and the point feature class was exported as a table to be imported in R. In R, each drought year VCI was rescaled to a range 0 to 100, where 0 mean low and 100 means high exposure. In order to rescale, the following equation was used:

$$Y = (X_i - X_{max})/(X_{min} - X_{max}) \qquad \textit{Eqn. 6}$$

Where, Y is the value of the variable when rescaled to the range 0 to 100, is the value of a variable before rescaling, X_{max} and X_{min} are the maximum and minimum value of a variable before rescaling, and X_i is a drought year VCI value.

After each drought year VCI value was rescaled, exposure was computed using the following formula (CIESIN, 2015):

$$E = (X_1a + X_2b + X_3c.... +X_zn)/n \qquad \textit{Eqn. 7}$$

Where E is exposure, X_1, X_2, X_3 and X_n are different VCI values of VCI for years, a, b, c, and z are weights of each drought year, and n is the number of drought years detected. All drought years were given equal weights.

Exposure based VCI (agricultural drought) was rescaled to 0 to 100, where 100 is the most exposed and 0 the least. Drought exposure was then compared with land cover provided by the Zambia Forestry Department.

2.4 Computation of SPI

The Standard Precipitation Index (SPI) was calculated using the SPI package in R. Monthly precipitation values of the wet season (October to April) for each station (from 1980 to 2015) were used with a predefined timescale of a 3 months moving window. Rainfall data was normalized using the gamma probability density function of the gamma distribution. The fitted function was used to calculate the cumulative distribution of the data points. Finally, the data points were transformed into standardised normal variates using the following equation:

$$SPI = (P_i - P_m)/\sigma \qquad \textit{Eqn. 8}$$

Where, P_i is the seasonal precipitation, P_m is the long-term mean, and σ is the standard deviation of the long-term record.

3. Results and Discussion

3.1 Relationship between Rainfall and NDVI

Correlation analysis between rainfall and NDVI was performed for all stations except Chipepo, whose coordinates were misplaced. A significant positive correlation was found to exist between rainfall and NDVI. The r and p values were 0.55 and 0.000095 respectively (**Figure 9.2**). Despite the existence of a relationship, it was fairly low. This was likely because of the different land cover types at each station, the gaps in the rainfall data and the low number of meteorological stations in the Southern Province.

The effect of land cover on the relationship between rainfall and NDVI correlation analysis was determined at each station. Correlation analysis was done at pixel level, where the pixel value at each station location was compared with the rainfall. The five stations were located in different land covers (**Table 9.1**). Land

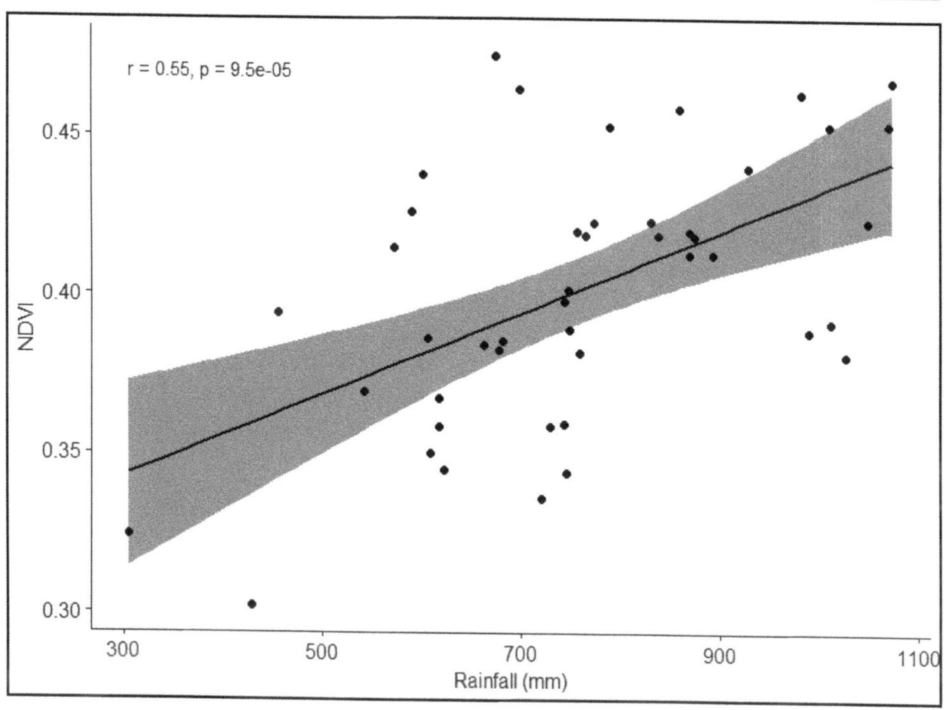

Figure 9.2: The Correlation between NDVI and Rainfall, Excluding Chipepo Station Data.

covers at the respective stations affected the relationship between rainfall and NDVI differently.

Table 9.1: Land Cover at each Meteorological Station

Meteorological Station	*Land Cover*
Chipepo	Inundated by Water
Choma	Built Up (Urban Area)
Kafue Folder	Commercial Cropland
Livingstone	Shrublands
Magoye	Bare ground

There were no significant relationships between rainfall and NDVI at the Chipepo, Choma and Kafue Folder stations. At the Chipepo station, the area was perennially inundated in water with no vegetation. Therefore, no significant relationship was expected to exist. The Kafue Folder meteorological station was located in an area covered by commercial crop land. Commercial croplands undergo irrigation throughout the year and plant biomass is not dependent on rainfall. Magoye meteorological station was located on bare grounds. No vegetation was present and, therefore, no correlation existed between NDVI and rainfall. Livingstone meteorological station was located in shrublands. A significant correlation existed between NDVI and rainfall at the station (r value was 0.61 and p was equal to 0.016). Moreover, the shrublands area responded positively to an increase in rainfall and negatively to a decrease. The characteristics of the land cover are that there short bushes spaced across plain covered by grass. Annual plants such as grass and shrubs respond quickly to changes in rainfall.

NDVI was at optimum level when rainfall is at 800 mm. However, NDVI begun to saturate when 800 mm of rainfall was exceeded. It remained at 0.46 as rainfall increases from 800 mm. The year 2005 experienced a decrease in NDVI. The lowest NDVI recorded was 0.1987 for the entire study period (**Figure 9.3**). Rainfall dropped to 650 mm in the same year. Hence, meteorological and agricultural droughts were experienced in 2005. It is possible that the effects of the agricultural drought that had occurred in the 2002 were still being experienced. The vegetation had not sufficiently recovered from water stress and the soil moisture had not been replenished completely.

Rainfall Anomaly (*RAF*) percentage showed that the region suffered droughts in the rainy seasons of 2001-02 and 2004-05. The effects of El-Niño in the 2014-15 and 2015-16 rainy seasons were also observed as negative rainfall anomaly percentages were observed. The most severe conditions were observed in the 2001-02 rainy season whilst the wettest conditions were observed in the 2005-06 season. NDVI corresponded with rainfall anomaly percentage as it dropped in the years 2002, 2005, 2015 and 2016, with all followed seasons that experienced drought. This was an indication that rainfall highly impacted vegetation.

In terms of spatial variation, the southern parts of the province received less rainfall than those in the northern parts (**Figure 9.4**). This was similar to the pattern of the two agro-ecological regions found in the province where the areas in the southern Region-I received less rainfall than Region-IIa in the north. However, the northern part of Kazungula did not follow this pattern and appears to receive more

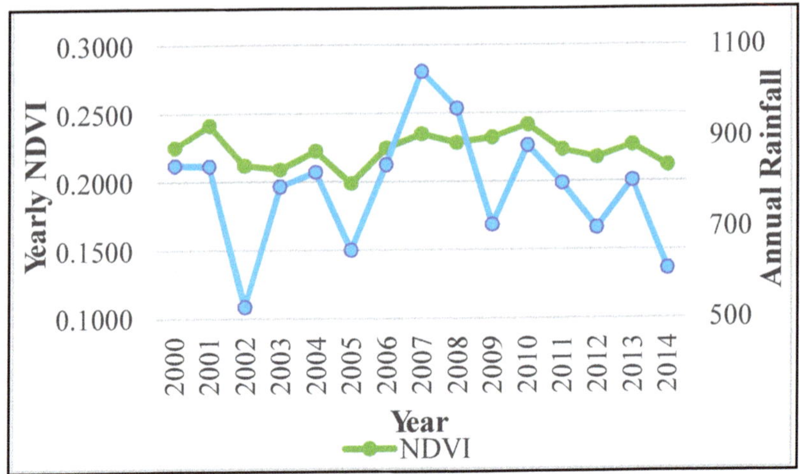

Figure 9.3: Trend of NDVI and Rainfall from 2000 to 2014 in Southern Province, Zambia.

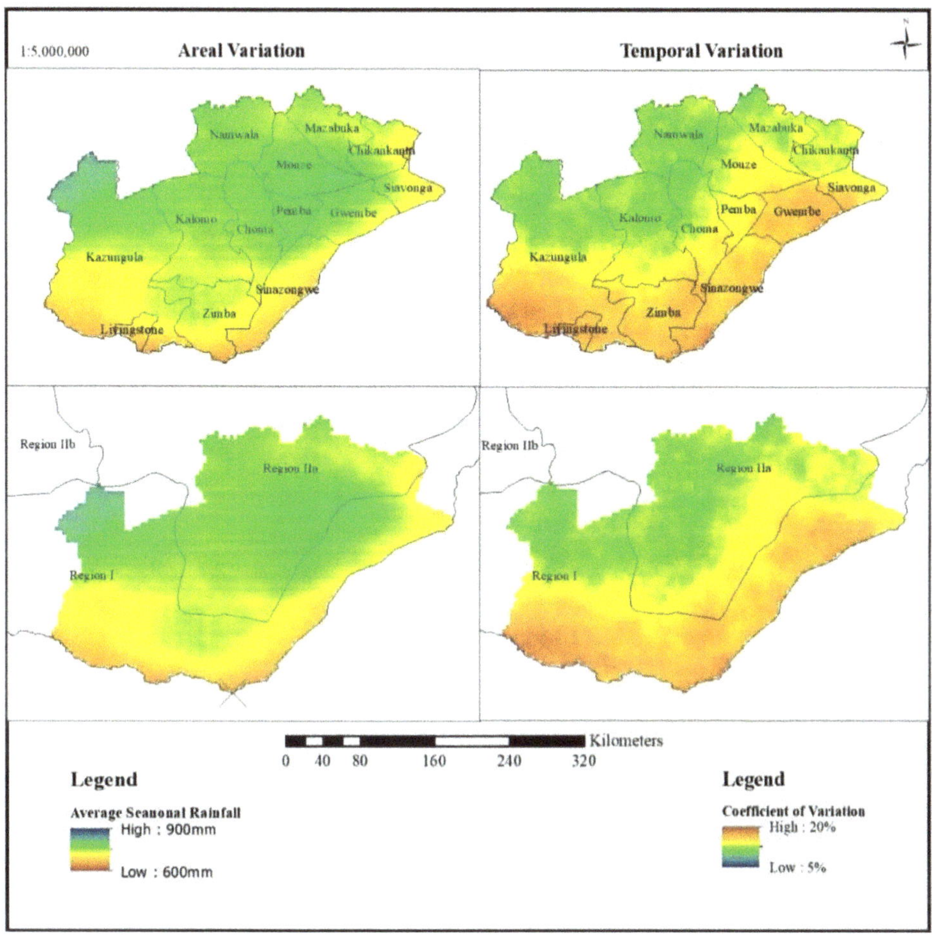

Figure 9.4: Average Seasonal Rainfall and Rainfall Variability.

rainfall, despite being in Region-I. Other districts within Region-I in the valley area experienced less rainfall.

Temporal variation was similar to spatial variation. The southern parts of the province experienced more rainfall variability from the long term seasonal average than those in the north. The variation in the south was as much as plus or minus 20 per cent, entailing less rainfall reliability than in the north. Following the two types of variability, the Southern Province has a north to south rainfall gradient, where rainfall decreases in both amount and reliability towards the south.

Similarly to rainfall anomaly percentage, drought was detected using VCI for the years 2002, 2005, 2015 and 2016. This is because agricultural drought was offset by prolonged meteorological drought in the Southern Province. However, agricultural drought exposure followed a different pattern to that of rainfall variation determined using TAMSAT rainfall estimates. Exposure seemed to follow a pattern

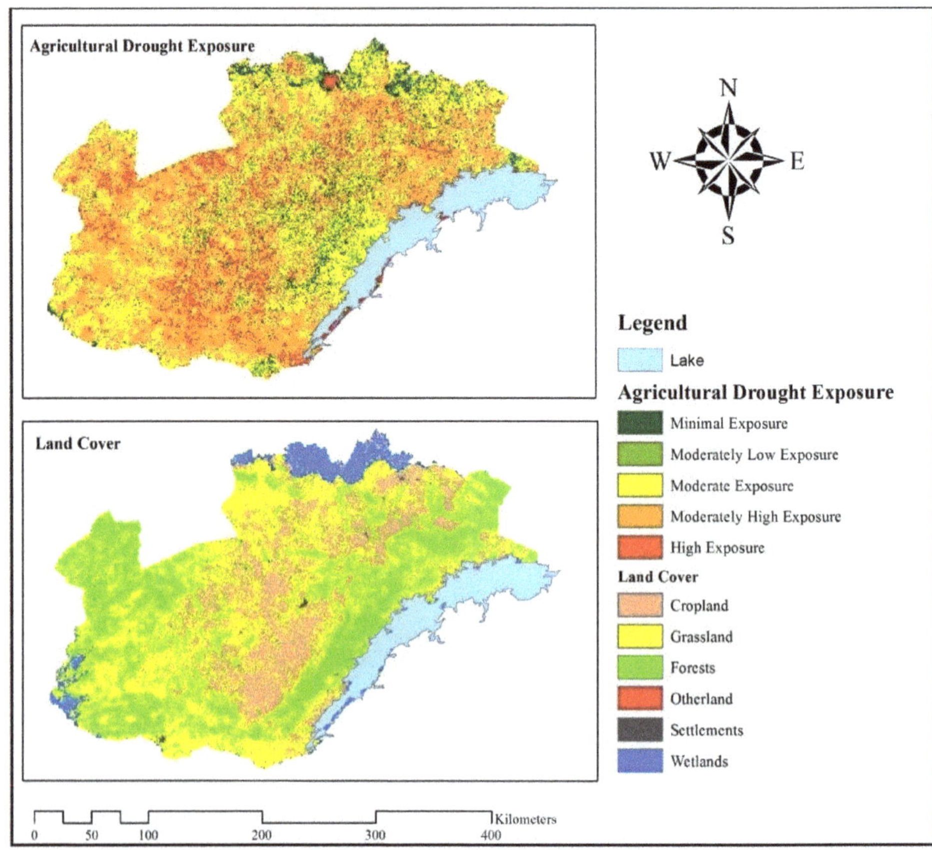

Figure 9.5: Agricultural Drought Exposure Based on VCI.

more similar to different land cover types than agro-ecological regions (**Figure 9.5**). Exposure was higher in areas covered by croplands, grasslands and forests, while areas in wetlands were less exposed.

Based on the 17 year study period (2000 to 2017), SPI shows a mild drought 27.45 per cent of the time, a moderate drought 5.60 per cent of the time and a severe drought 2.52 per cent of the time (**Table 9.2**). These percentages are expected because SPI is standardized (WMO, 2012) and it allows the SPI to determine the frequency of a current drought as well as the probability of the rainfall necessary to end it (Mckee, *et al.*, 1993). A mild drought is expected to occur once in every 4 years, a moderate drought once in 18 years and a severe drought once in every 40 years. No extreme drought events were recorded within the 17 year study period.

4. Conclusion

A significant correlation exists between NDVI and rainfall. Therefore, NDVI can be used as proxy to rainfall when studying drought. However, the relationship

Table 9.2: Drought Event Severity

Category	*Number of Occurrences in 17 Years*	*Relative Frequency (per cent)*	*Total Length of Time Experienced in 17 Years (Years)*	*Drought Event Severity*
Extremely Dry	0	0.00	0	N/A
Severely Dry	9	2.52	0.43	40
Moderately Dry	20	5.60	0.95	18
Mildly Dry	98	27.45	4.67	4
Normal	165	46.22	7.86	2
Moderately Wet	42	11.76	2.00	9
Very Wet	18	5.04	0.86	20
Extremely Wet	5	1.40	0.24	71
Totals	**357**	**100.00**	**17**	**1**

is affected by land cover. Different covers display different behaviors during times of drought. For instance, NDVI remains high in wetland areas despite experiencing drought. The years 2002, 2005 and 2014 experienced low rainfall and drought was detected using VCI and rainfall anomaly percentage. The year 2002 had the lowest rainfall. However, the lowest NDVI was observed in 2005. NDVI also displayed a lag in responding to moisture deficit. The Southern Province displayed a north to south gradient in rainfall variability, where variability increased towards the south, a pattern similar to the agro-ecological zones. However, drought exposure based on VCI displayed a pattern that followed land cover.

Acknowledgements

We would like to acknowledge the National Science and Technology Council (NSTC) of Zambia for providing the financial assistance for this study. We would also like to appreciate the help and inputs from the South African National Space Agency (SANSA) and the University of Zambia. Most of all, we like to recognize Dr. Augustine Mulolwa for initiating the project.

References

1. Amalo, L. F., Hidayat, R. and Haris, 2017. Comparison between remote-sensing-based drought indices in East Java. *IOP Conf. Series: Earth and Environmental Science,* Volume 54, pp. 1-7.
2. Bhuiyan, C., 2004. Various Drought Indices For Monitoring Drought Condition in Aravalli Terrain of India. *ISPRS International Journal of Geo-Information.*
3. Camberlin, P., Martiny, N., Philippon, N. and Richard, Y., 2007. Determinants of the Interannual Relationships Between Remote Sensed Photosynthetic Activity and Rainfall in Tropical Africa. *Remote Sensing of Environment,* Volume 106, pp. 199-216.

4. Cap-Net, 2015. *Drought risk reduction training manual.* New York: United Nations Develoment Programme.

5. Chen, S. *et al.*, 2015. Temperature Vegetation Dryness Index Estimation of Soil Moisture under Different Tree Species. *Sustainability,* Volume 7, pp. 11401-11417.

6. Chopra, P., 2006. *Drought Risk Assessment using Remote Sensing and GIS: A case study of Gujarat.,* Enschede,: University of Twente.

7. CIESIN, 2015. *Step-by-Step Guide to Vulnerability Hotspots Mapping: Implementing the Spatial Index Approach.* New York: Columbia Umiversity.

8. Congalton, R. G., 1991. A Review of Assessing the Accuracy of Classifications of Remotely Sensed Data. *Remote Sensing of Environment,* 37(1), pp. 35-46.

9. CSO, 2011. *Zambia 2010 Census of Population and Housing: Preliminary Population Figures.,* Lusaka: Government Printers.

10. CSO, 2012. *Zambia 2010 Census of Population and Housing: Population Summary Report.* Lusaka: Central Statistics Office (CSO).

11. CSO, 2015. *Agriculture Statistics - Zambia Data Portal.* [Online]

    Available at: http://zambia.opendataforafrica.org/ionawve/agriculture-statistics

    [Accessed 20 September 2017].

12. Cunningham, M., 2009. *More than Just the Kappa Coefficient: A Program to Fully Characterize Inter-Rater Reliability between Two Raters.* Washingtone D.C: SAS Institute.

13. Damizadeh, M., Bahram, S. and Gieske, A., 2001. *Studying Vegetation Responses and Rainfall Relationship Based-On NOAA/AVHRR Images.* Singapore, Centre for Remote Imaging, Sensing and Processing (CRISP).

14. Davenport, M. L. and Nicholson, S. E., 1993. On the Relation Between Rainfall and the Normalized Difference Vegetation Index for Diverse Vegetation Types in East Africa. *International Journal of Remote Sensing,* 14(12), pp. 2369-2389.

15. Day, M. and Saive, J., 2009. Testing Shelford's Law of Tolerance in Amphipods (Gammarus Minus) Exposed to Varying PH and Conductivity. *Journal of Ecological Research,* Volume 9, pp. 40-44.

16. Doesken, N. J., Mckee, T. B. and Kleist, J., 1991. *Development of a Surface Water supply Index for the Western United States,* Fort Collins: Colorado State University.

17. ESRI, 2007. *How Weighted Overlay works.* [Online]

    Available at: http://edndoc.esri.com/arcobjects/9.2/net/shared/geoprocessing/spatial_analyst_tools/how_weighted_overlay_works.htm

    [Accessed 11 September 2017].

18. ESRI, 2016. *Weighted Overlay—Help | ArcGIS for Desktop.* [Online]

    Available at: http://desktop.arcgis.com/en/arcmap/10.3/tools/spatial-analyst-toolbox/weighted-overlay.htm

    [Accessed 11 September 2017].

19. ESRI, 2017. *How Inverse Distance Weighted Interpolation Works.* [Online]

    Available at: http://pro.arcgis.com/en/pro-app/help/analysis/geostatistical-analyst/how-inverse-distance-weighted-interpolation-works.htm

    [Accessed 10th October 2017].

20. FAO, 2016. *Southern Africa El Niño Response Plan (2016/17).* Rome: Food and Agricultural Organisation of the United Nations.

21. FAO, 2017. *CHAPTER 2: CROP WATER NEEDS.* [Online]

    Available at: http://www.fao.org/docrep/s2022e/s2022e02.htm

    [Accessed 14 August 2017].

22. FEWS NET, 2007. *Products: Croplands Water Requirement Satisfaction Index (WRSI).* [Online]

    Available at: https://earlywarning.usgs.gov/fews/product/128

    [Accessed 5 October 2017].

23. FEWS NET, 2014. *Zambia Livelihood Zones and Descriptions.* Lusaka: USAID.

24. FEWS NET, 2017. *Early Warning and Environmental Monitoring Program.* [Online]

    Available at: https://earlywarning.usgs.gov/fews

    [Accessed 6th October 2017].

25. Ganesh, S., 2013. *Investigation of the Utility of the Vegetation Condition Index (VCI) As an Indicator of Drought,* College Station: Texas A and M University.

26. Gao, Z., Gao, W. and Chang, N.-B., 2010. Integrating Temperature Vegetation Dryness Index (TVDI) and Regional Water Stress Index (RWSI) for Drought Assessment with the Aid of LANDSAT TM/ETM+ Images. *International Journal of Applied Earth Observation and Geoinformation,* 13(3), pp. 495-503.

27. Guenang, G. M. and Kamga, F. M., 2014. Computation of the Standardized Precipitation Index (SPI) and Its Use to Assess Drought Occurrences in Cameroon over Recent Decades. *Journal of Applied Meteorology and Climatology,* Volume 53, pp. 2310-2324.

28. Hall, G., 2015. [Online]

    Available at: http://www.hep.ph.ic.ac.uk/~hallg/UG_2015/Pearsons.pdf

    [Accessed 13 December 2016].

29. Hauke, J. and Kossowski, T., 2011. Comparison of Values of Pearson's and Spearman's Correlation Coefficients on the Same Sets of Data. *Quaestions Geagraphicae,* 30(2), pp. 87-93.

30. Huisman, O. and de By, R. A., 2009. *Principles of Geographic Information Systems: An introductory textbook.* Fourth ed. Enschede: International Institute for Geo-Information Science and Earth Observation (ITC).

31. IDMP, 2017. *Deciles.* [Online]

    Available at: http://www.droughtmanagement.info/deciles/

    [Accessed 28th September 2017].

32. Kogan, F. N., 1997. Global Drought Watch from Space. *Bulletin of the American Meteorological Society,* 78(4), pp. 621-636.

33. Kogan, F. N., 2001. Operational Space Technology for Global Vegetation Assessment. *Bulletin of the American Meteorological Society,* 82(9), pp. 1949-1964.

34. Kundzewicz, Z. W. and Matczak, P., 2015. *Hydrological extremes and security.* Göttingen, Copernicus Publishers, pp. 44-53.

35. Latif, M. S., 2014. Land Surface Temperature Retrival of Landsat-8 Data Using Split Window Algorithm - A Case Study of Ranchi District. *Internal Journal of Engineering Development and Research,* 2(4), pp. 3840-3849.

36. Lekprichakul, T., 2008. *Impact of 2004/2005 Drought on Zambia's Agricultural Production.* Kyoto: Research Institute for Humanity and Nature (RIHN).

37. Maidment, R. I. *et al.,* 2014. The 30 year TAMSAT African Rainfall Climatology And Time series (TARCAT) data set. *Journal of Geophysical Research,* 22 September, 119(18), pp. 10619-10644.

38. Maidment, R. I. *et al.,* 2017. A New, Long-term Daily Satellite-Based Rainfall Dataset for Operational Monitoring in Africa. *Scientific Data,* Volume 4, pp. 1-17.

39. Makaudze, E. M. and Miranda, M., 2008. *Using Remotely-Sensed Vegetation Condition Index to investigate the feasibility of Drought Insurance for Smallholder Farmers in Zimbabwe.* Cape Town, University of Western Cape.

40. Mckee, T. B., Doesken, N. J. and Kleist, J., 1993. The Relationship of Drought Frequency and Duration to Time Scales. *AMS 8th Conference on Applied Climatology,* pp. 179-184.

41. Meng, L. *et al.,* 2008. The Calculation of TVDI Based on the Composite Time of Pixel and Drought Analysis. *The International Archives of the Photogrammetry, Remote Sensing and Spatial Information Sciences,* 38(2), pp. 519-524.

42. MESA SADC-THEMA, 2014. *Drought Service Manual.* [Online]

    Available at: http://www.mesasadc.org/sites/default/files/manuals/Service%20Product%20Manual%20-%20Drought.pdf

    [Accessed 26th September 20117].

43. MESA SADC-THEMA, 2017. *Drought.* [Online]

    Available at: http://www.mesasadc.org/drought

    [Accessed 28th September 2017].

44. Meshu, T. and Berhan, G., 2013. Interactive Drought Monitoring Information System: A Case Study on Standardized Precipitation Index for Ethiopia. *HiLCoE Journal of Computer Science and Technology,* 1(2), pp. 125-131.

45. Mishra, M. M. and Nagarajan, R., 2013. Hydrological Drought Assessment in Tel River Basin Using Standardized Water Level Index (SWI) and GIS Based Interpolation Techniques. *International Journal of Conceptions on Mechanical and Civil Engineering,* 1(1), pp. 3-6.

46. Mu, Q. *et al.,* 2013. A Remotely Sensed Global Terrestrial Drought Severity Index. *Bulletin of American Meteorological Society,* 94(1), pp. 83-98.

47. Namias, J., 1983. Some causes of United States drought. *Journal of Climate and Applied Meteorology,* 22(1), pp. 30-39.

48. Narasimhan, B. and Srinivasan, R., 2005. Development and Evaluation of Soil Moisture Deficit Index (SMDI) and Evapotranspiration Deficit Index (ETDI) for Agricultural Drought Monitoring. *Agricultural and Forest Meteorology,* 133(1-4), pp. 69-68.

49. National Weather Service, 2006. *Drought: Public Fact Sheet.* s.l.:s.n.

50. National Weather Service, 2012. *Drought Fact Sheet.* Silver Spring: National Oceanic and Atmospheric Administration.

51. NDMC, 2006. *Types of Drought.* [Online]

    Available at: http://drought.unl.edu/DroughtBasics/TypesofDrought.aspx

    [Accessed 22 August 2016].

52. NIST, 2016. *1.3.6.6.11. Gamma Distribution.* [Online]

    Available at: http://www.itl.nist.gov/div898/handbook/eda/section3/eda366b.htm

    [Accessed 10 March 2017].

53. Owrangi, M. A. *et al.,* 2011. Drought Monitoring Methodology Based on AVHRR Images and SPOT Vegetation Maps. *Journal of Water Resource and Protection,* Volume 3, pp. 325-334.

54. Palmer, W. C., 1965. *Meteorological Drought,* Washinton D.C: US Weather Bureau.

55. Phyisics World, 2017. *Rain Drains Energy from the Atmosphere.* [Online]

    Available at: http://physicsworld.com/cws/article/news/2012/feb/24/rain-drains-energy-from-the-atmosphere

    [Accessed 13th November 2017].

56. RisCura, 2017. *The effect of El Niño on key Southern African countries.* [Online]

    Available at: http://www.riscura.com/brightafrica/el-nino/effect-southern-african-countries/

    [Accessed 13th November 2017].

57. Rousvel, S. *et al.*, 2013. Comparison between Vegetation and Rainfall of Bioclimatic Ecoregions in Central Africa. *Atmosphere,* Volume 4, pp. 411-427.

58. Sandholt, I., Rasmussen, K. and Andersen, J., 2002. A simple Interpretation of the Surface Temperature/Vegetation Index Space for Assessment of Surface Moisture Status. *Remote Sensing of Environment,* Volume 79, pp. 201-224.

59. Senay, G. B. and Verdin, J., 2002. *Evaluating the Performance of Crop Water Balance Model in Estimating Regional Crop Production.* Pecora, ISPRS Commission.

60. Szep, I. J., Janos, M. and Dunkel, Z., 2005. Palmer Drought Severity Index as Soil Moisture Indicator: Physical Interpretation, Statistical Behaviour and Relation to Global Climate. *Physics and Chemistry of the Earth,* 30(1-3), pp. 231-243.

61. TAMSAT, 2017. *About Us.* [Online]

    Available at: https://www.tamsat.org.uk/

    [Accessed 5th October 2017].

62. Tarnavsky, E. *et al.*, 2014. Extension of the TAMSAT Satellite-based Rainfall Monitoring over Africa and from 1983 to Present. *Journal of Applied Meteorology and Climate,* 21 August, 53(12), pp. 2805-2822.

63. Tempfli, K., Kerle, N., Huurneman, G. C. and Janssen, L. L., 2009. *Principles of Remote Sensing: An introductory textbook.* Fourth ed. Enschede: International Institute for Geo-Information Science and Earth Observation (ITC).

64. Thenkabail, P. S., Gamage, M. S. and Smakhtin, V. U., 2004. *The Use of Remote-Sensing Data for Drought Assessment and Monitoring in Southwest Asia,* Colombo: International Water Management Institute.

65. U.N.I.S, 2007. *Drought Risk Reduction Framework and Practices: Contribution to the Implementation of Hyogo Framework.* s.l.:ISDR.

66. Unganai, L. S. and Kogan, F. N., 1998. Drought Monitoring and Corn Yield Estimation in Southern Africa From AVHRR Data. *Remote Sensing of Environment,* 63(3), pp. 219-232.

67. UNOCHA, 2017. *El Niño in Southern Africa.* [Online]

    Available at: http://www.unocha.org/legacy/el-nino-southern-africa

    [Accessed 13th November 2017].

68. USGS, 2015. *Using the USGS Landsat 8 Product.* [Online]

    Available at: https://landsat.usgs.gov/using-usgs-landsat-8-product

    [Accessed 21 January 2016].

69. Venables, W. N., Smith, D. M. and R Core Team, 2018. *An Introduction to R:.* Los Angelas: R Core Team.

70. Verhulst, N. and Govaerts, B., 2010. *The Normalized Difference Vegetation I(NDVI) GreenSeekerTM Handheld Sensor: Toward the Integrated Evaluation of Crop Management. Part A: Concepts and Case Studies.* Mexico: CIMMYT.

71. Wang, J., Rich, P. M. and Price, K. P., 2003. Temporal Responses of NDVI to Precipitation and Temperature in the Central Great Plains, USA. *International Journal of Remote Sensing,* 24(11), pp. 2345-2364.
72. WMO, 2012. *Standardized Precipitation Index User Guide (WMO-No. 1090).* Geneva: World Meteorological Organisation (WMO).
73. Zoungrana, B. J. *et al.,* 2015. Land Use/Cover Response to Rainfall Variability: A Comparing Analysis between NDVI and EVI in the Southwest of Burkina Faso. *Climate,* Volume 3, pp. 63-77.

Chapter 10

Impact of Extreme Climate on Crop Production and Management Techniques in Batticaloa District, Sri Lanka: Review on Flood and Drought

A. Narmilan[1], B.G.N. Sewwandi[1] and S. Puvanitha[2]*

[1]*Department of Biosystems Technology, Faculty of Technology, South Eastern University of Sri Lanka*
[2]*Department of Biosystems Technology, Faculty of Technology, Eastern University, Sri Lanka*
**E-mail: narmilan@seu.ac.lk*

ABSTRACT

Climate change is one of the most widely researched and discussed topical problems affecting many sectors specially agriculture. Sri Lanka's economy is heavily dependent on agriculture which employs a large percentage of the population. Developing countries are more vulnerable to climate change, as they will be the most affected, but have the minimum capacity to adapt to it. Climate change may be due to natural internal processes, external forcing or persistent anthropogenic changes in the composition of the atmosphere or in land use. Impacts of climate variability and change on the agricultural sector are projected to steadily manifest directly from changes in land and water regimes, the likely primary conduits of change. Climate change thus represents an additional burden that for farmers translates into production risks associated with crop yields, probabilities of extreme events, timing of field operations, and timing of investments in new technologies. Long term shifts in rainfall and temperature regimes, extreme climatic events such as droughts and flood can cause substantial damage to crops. In order to mitigate the impact of flood and drought, appropriate management measures have to be implemented. These measures can be classified into structural measures, non-structural measures and integration of structural and non-structural measures. Structural Measures include protection of the vulnerable area up to a certain level of extreme event and preferred by engineers and local people. Non-structural measures focus on reduction of loss or damage which referred by social scientists and conservationist. Integration of structural and non-structural measures is essential for effective disaster management. This review was conducted to study the impact of climate change on crop production and the appropriate measures which uses science and technology in disaster management. Therefore, the main objective is to identify the climate change hazards and possible disaster management projects in Batticaloa and to establish the broad agreement of the Batticaloa scientific community on critical issues related to the climate change hazards.

Keywords: *Climate change, Crop production, Disaster management, Extreme weather, Flood, Drought, Batticaloa.*

1. Introduction

Climate change has been defined as a change of climate which is attributed directly or indirectly to human activity that alters the composition of the global atmosphere in addition to natural climate variability observed over comparable time periods (Obioha, 2009). The agricultural sector of Sri Lanka, like in most developing nations, is crop based and is the primary source of employment to the rural populations (Devendra, 2002). Availability of research data on issues of climate change in Batticaloa District is limited. Lack of such information is also known to be an obstacle for strategic planning, policy development and priority setting in Sri Lanka. Therefore, the objectives of this review is to identify the climate change hazards and possible disaster management projects in Batticaloa and to establish the broad agreement of the Batticaloa scientific community on critical issues related to the climate change hazards.

2. Discussions

2.1 Impact of Climate Change on Crop Production

Agriculture, especially crop production, is highly sensitive to both short and long-term changes in climate. Temperature increases will reduce the durations of most cereal crops by hastening their phonological development (Wheeler *et al.*, 2000). Rice is highly influenced by variations in temperature and rainfall (Fernando 2000). A study by Cheng *et al.* (2009) has shown that increasing night temperatures will reduce the stimulatory effects of increasing CO_2 on rice yields. Grain development of rice is highly sensitive to temperature during its reproductive stage, with significant increases in grain sterility occurring when temperature increases beyond 34 °C even for a few hours (Horie *et al.*, 2000). In addition to the long term shifts in rainfall and temperature regimes, extreme climatic events such as droughts and episodes of high temperature can cause substantial damage to both rice and tea crops. Such damage has serious repercussions on the national food security and foreign exchange earnings while causing substantial socio-economic damage to the large number of families that are dependent on these two crops. Similar adverse impacts are experienced by other crops (Porter and Semenov 2005).

2.2 Statistical Information on Extreme Weather and their Effects on Crop Production in Batticaloa District

Sri Lanka statistical abstract reported that analysis of air temperature and rainfall data collected by the Department of Meteorology over a period of more than 100 years have shown an increasing trend in the annual mean air temperature over the entire Island, particularly during the more recent period, 1961-1990. This increase was found to be approximately 0.16°C per decade. Batticaloa is ranked among the five districts which recorded the lowest productivity in paddy farming in the 2010/11 'Maha' season. Certainly, it may be argued that this is due in part to the severe floods experienced in this region in early 2011 (Department of Census and Statistics, 2016). The Dry and Intermediate zones are the most vulnerable to drought, with the districts of Jaffna, Killinochchi, Batticaloa, Polonnaruwa, Anuradhapura and Kurunegala having the highest probability of experiencing drought (Ministry of Environment, 2011).

Paddy irrigation requirements are predicted to increase on average across Sri Lanka by 23 per cent. The highest proportional increase is predicted to occur in Batticaloa, by 45 per cent. This increase is mainly due to the decreased rainfall during January and February months. In the recent data records, the largest downward trend of 11.16 mm/year was observed at Batticaloa. The highest decrease is predicted in Trincomalee and Batticaloa as 27 and 29 per cent, respectively creating more drought problems in these areas (Premalal 2009). Disaster Situation Report (2017) reported that 24,129 families affected due drought in Batticaloa district in 2017. Flood Situation Report (2014) reported that, 124,071 families affected due flood in Batticaloa district in 2017, 3,361 houses fully damaged and 6,736 houses partially damaged due to flood in 2014. The highest number of persons (>3,604,769 people) affected by disaster is recorded in the Batticaloa district during the period of 1974-2006. The range between 8000-90550 people affected due to drought in the Batticaloa district in the period of 1974-2006. More than 2,394,704 people affected due to flood in the Batticaloa district in the period of 1974-2006. The agricultural land area of 343 to 3530ha are affected due to drought in the Batticaloa district in the period of 1974-2006. The agricultural land area of 21,019 to 36,583ha affected due to flood in the Batticaloa district in the period of 1974-2008 (Disaster Information Management system in Sri Lanka, 2017).

Disaster Profile of Sri Lanka (2012) reported that cost of agricultural damages and losses is 15,070 LKR million in Batticaloa, Polonnaruwa, Anuradhapura, and Ampara districts due to flood in 2011.

2.3 Disaster Management Projects Implemented in Batticaloa District

Climate Resilient Action Plans of Coastal Areas of Sri Lanka (CCSL) implemented awareness programs with the school heads and teachers in Batticaloa district with the help of Batticaloa municipal council (BMC). Following themes have been discussed at the awareness program:

1. Post disaster health hazards (Mosquito borne diseases, Water borne diseases)

2. Use of trees/plants for climate change adaptation
3. Preparation for disaster (Cyclones, floods, tsunami)
4. Safe home/school for disasters
5. Save energy at home/school

Disaster Management Centre (DMC) has implemented following measures to reduce negative impact of any disaster (DMC-NDMCC, 2015)-

1. Risk awareness and assessment (hazard mapping, vulnerability assessment, capacity building).
2. Knowledge development (public education, training on community based disaster management, information sharing)
3. Application of measures (environmental protection measures, protection of critical places of public places)
4. Early warning systems (forecasting, dissemination of warnings, preparedness measures, response capacities).

Project team of CCSL developed a project to provide recommendations to Sri Lankan cities to develop action plans to build disaster resilient cities based on the lessons learnt in the implementation of different project activities. The four Climate Resilient Adaptation Strategies and Supporting Action Plans (CRASSAPs) under this project are (CRASSAP, 2010):

1. Water resource management, especially drainage and sanitation impact from more intense rainfall events, supported by a Pre-Feasibility Study (PFS) with funding support from the SIDA (Swedish International Development Cooperation Agency) supported City Development Initiative for Asia (CDIA).
2. A multi-purpose green belt (12 km in length) was established to protect the lagoon and coastal areas, restore mangrove eco-systems and coastal bio-diversity in BMC area.
3. GIS-based Rapid Response Systems (RRS) and two Knowledge Management centers for climate exacerbated disasters were established at BMC and Negombo Municipal Council (NMC) with a training (1 month for selected 50 participants) and equipment (building, Software, 20 computers, 10 GPS, printers, scanners, broadband facilities).
4. Disaster resilient; energy efficient; low-cost shelter adaptation training, supported by local resource based- livelihood diversification options for 100 participants live in vulnerable areas.

2.3 Recommended Disaster Management Technology

1. ***GIS and remote sensing:*** GIS models having low cost and simple data requirement are likely to attract the local authorities in the developing countries to adopt this technology as an essential input towards a comprehensive flood management system (Saiful Islam *et al.*, 2012).

2. ***Internet:*** Launching of a well-defined website is a very cost-effective means of making an intra-national and international presence felt. It provides a new and potentially revolutionary option for the rapid, automatic and global dissemination of disaster information.
3. ***Satellite Communication:*** Satellite data can be effectively used for mapping and monitoring the flood inundated areas, flood damage assessment, flood hazard zoning and post-flood survey of rivers configuration and protection works
4. ***Flood forecasting system:*** The flood forecasting and warning system is used for alerting the likely damage centers well in advance of the actual arrival of floods, to enable the people to move and also to remove the moveable property to safer places or to raised platforms specially constructed for the purpose.

3. Conclusions and Recommendations

Adverse weather and climatic changes, capital shortages, high cost of inputs, access to credit difficulties and poor quality of output are affecting agricultural productivity and eventually impact on the incomes of the farmers in Batticaloa district. However, adequate and relevant information about impact of climate change in Batticaloa district is generally lacking. Therefore, management practices and cultivars will probably have to be adjusted to maintain agricultural crop production under a changing climate. Promoting local and international research collaborations among the researchers who worked in the same field, exchanging and implementing the knowledge generated to address the similar issues will overcome the problems mentioned above. The advancement in Information Technology in the form of Internet, GIS, Remote Sensing, Satellite communication, *etc.* can help a great deal in planning and implementation of hazards reduction. New technologies for public communication should be made available to use and natural disaster mitigation messages should be conveyed through these measures for maximum benefit. In conclusion, climate change will decrease crop yields in the long term, unless one slows climate change and/or adapts new management practices and improved cultivars.

References

1. Cheng W., Sakai H., Yagi K. and Hasegawa T. 2009. Interactions of elevated CO_2 and night temperature on rice growth and yield. *Agricultural and Forest Meteorology* 149: 51-58
2. CRASSAP 2010. Climate resilient action plans for Coastal urban areas in Sri Lanka (CCSL) (Activity 1.3 - Activity 1.5). *Lessons Learnt Training Manual* for Sri Lankan Coastal Cities.
3. Department of Census and Statistics 2016. *Statistical Abstract*. Sri Lanka.
4. Devendra C. 2002. Crop-animal systems in Asia: future perspectives. *Agricultural Systems* 71: 179 -186

5. Disaster Information Management System in Sri Lanka. 2017. *Desinventar.lk.* Retrieved 27 November 2017, from http://www.desinventar.lk.

6. DMC-NDMCC (National Disaster Management Coordination Committee) 2015. Risk reduction awareness: http://www.dmc.gov.lk/NDMCC/Task%20of%20 NDMCC.htm. Accessed on 10 Nov. 2017.

7. Fernando T.K. 2000. Impact of climate change on paddy production in Sri Lanka. *Global Environmental Research* 2: 169-176

8. Horie T., Baker J.T., Nakagawa H., Matsui T. and Kim H.Y. 2000. Crop ecosystem responses to climatic change: Rice. In: Climate Change and Global Crop Productivity. (Eds. K.R. Reddy and H.F. Hodges) pp. 81-106. CAB International, Wallingford, UK.

9. Ministry of Environment in Sri Lanka. 2011. National Climate Change Adaptation for Sri Lanka 2011 to 2016. Colombo, Sri Lanka. DOI: http://dx.doi.org/10.4038/jnsfsr.v38i2.2032. Accessed on (8 Nov. 2017).

10. Obioha E. 2008. Climate change, population drift and violent conflict over land resources in north eastern Nigeria. *Journal of Human Ecology* 23: 311-324

11. Porter J.R. and Semenov M.A. 2005. Crop responses to climatic variation. *Philosophical Transactions of the Royal Society B: Biological Sciences* 360: 2021–2035

12. Premalal K.H.M.S. 2009. Climate change in Sri Lanka. *Proceedings of Global Climate Change and its Impacts on Agriculture, Forestry and Water in the Tropics.* Kandy, Sri Lanka.

13. Saiful Islam, Sabir Khan and R.K.Sharma. 2012. Role of recent technology in disaster management. *IJRREST: International Journal of Research Review in Engineering Science and Technology* 1: 35-40

14. Wheeler T.R., Crauford P. Q., Ellis R.H., Porter J.R. and Vara Prasad P.V. 2000. Temperature variability and the yield of annual crops. *Agriculture, Ecosystems and Environment* 82: 159-167

APPENDICES

People Affected Due to Disaster-Spatial Distribution: 1974-2006

Figure 10.1a Shows spatial distribution of people affected map illustrates the geographical distribution of people affected by disaster across the districts in the country. The highest number of persons affected by disaster is recorded in the Batticaloa district (more than 3,604,769 people). *Source: www.desinventar.lk*

People Affected Due to Drought-Spatial Distribution: 1974-2006

Figure 10.1b shows that spatial distribution of people affected map illustrates the geographical distribution of people affected by drought across the districts in the country. The range between 8000-90550 people affected due to drought in the Batticaloa district in period of 1974-2006. *Source: www.desinventar.lk.*

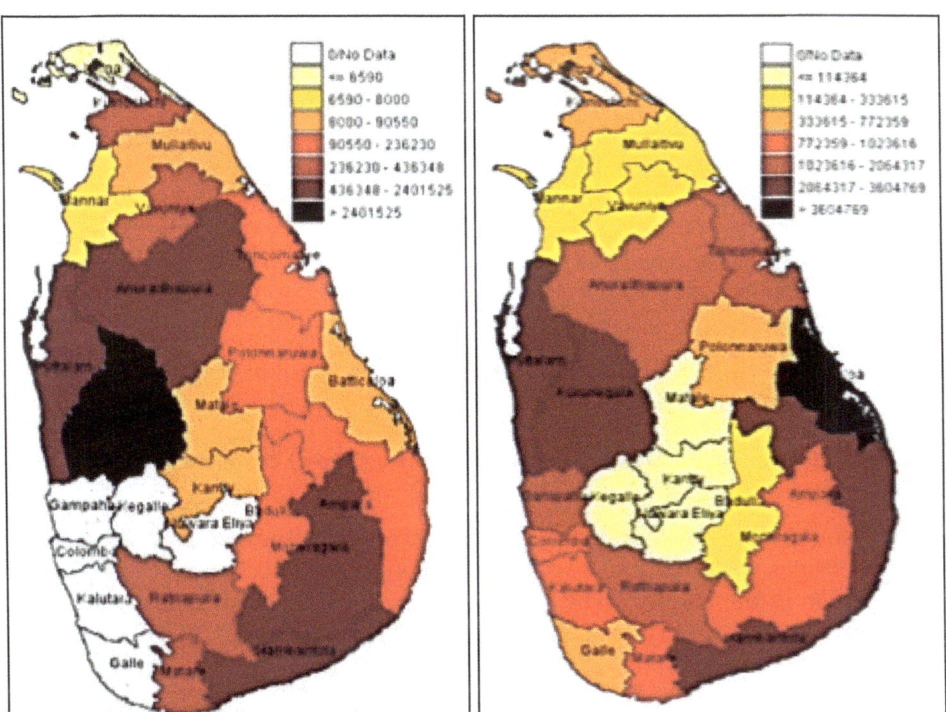

Figure 10.1a: People Affected Due to Disaster-Spatial Distribution: 1974-2006.

Figure 10.1b: People Affected Due to Drought-Spatial Distribution: 1974-2006.

People Affected Due to Flood-Spatial Distribution: 1974-2006

Figure 10.2a shows that spatial distribution of people affected map illustrates the geographical distribution of people affected by disaster across the districts in the

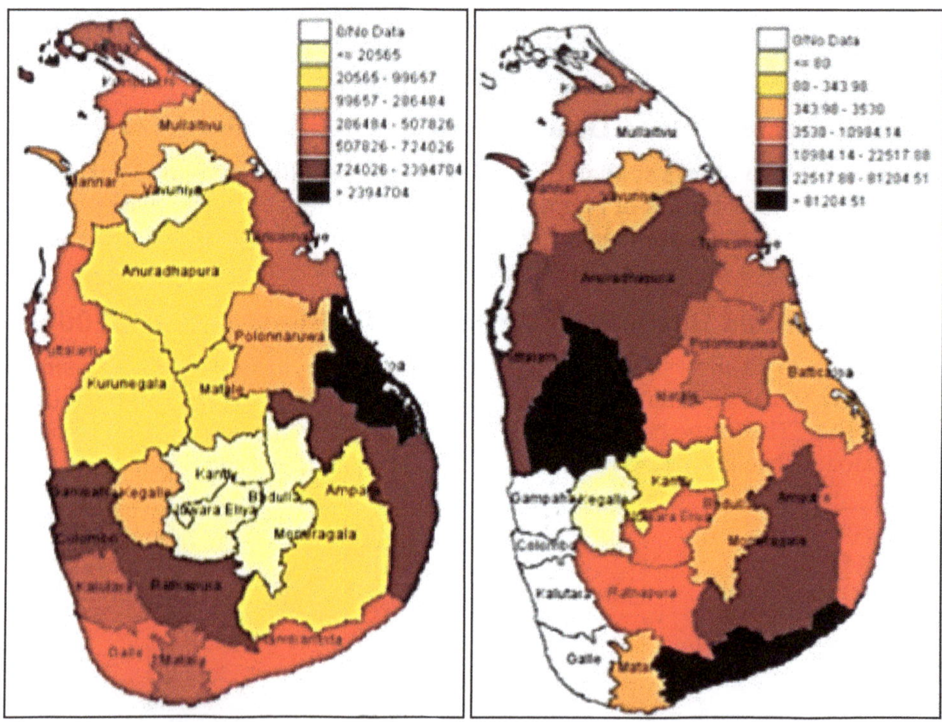

Figure 10.2a: People Affected Due to Flood-Spatial Distribution: 1974-2006.

Figure 10.2b: Agricultural Losses Due to Drought (in ha) Spatial Distribution: 1974-2006.

country. More than 2,394,704 people affected due to flood in the Batticaloa district in period of 1974-2006. Source: *www.desinventar.lk.*

Agricultural Losses Due to Drought (in Hectare) Spatial Distribution: 1974-2006

Figure 10.2b shows that spatial distribution of agricultural losses due to drought map illustrates the geographical distribution of people affected by disaster across the districts in the country. The range between 343–3530ha of agricultural land affected due to drought in the Batticaloa district in period of 1974-2006. *Source: www.desinventar.lk*

Agricultural Losses Due to Flood (in Hectare) Spatial Distribution: 1974-2008

Figure 10.3 shows that spatial distribution of agricultural losses due to flood map illustrates the geographical distribution of people affected by disaster across the Districts in the country. The range between 21,019 – 36,583ha of agricultural land affected due to flood in the Batticaloa district in period of 1974-2008. *Source: www.desinventar.lk.*

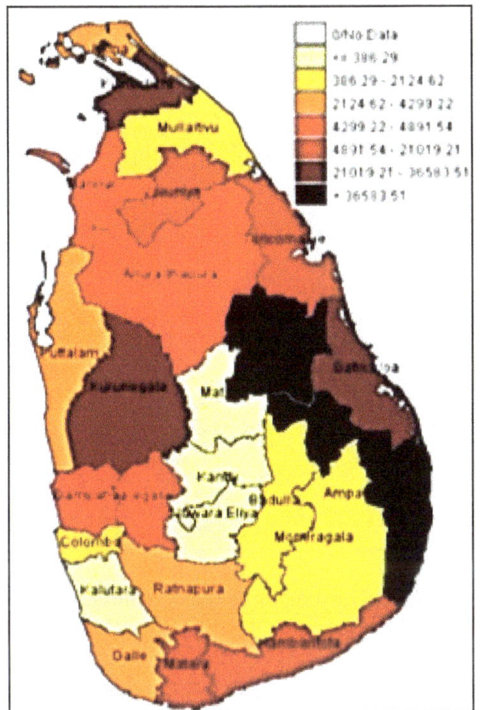

Figure 10.3: Agricultural Losses Due to Flood (in ha) Spatial Distribution: 1974-2008.

Figure 10.4a shows the cumulative number of people affected by flood during 2002-12 in Batticaloa district. It shows highest number of people affected by flood. *Source: Disaster Profile of Sri Lanka.*

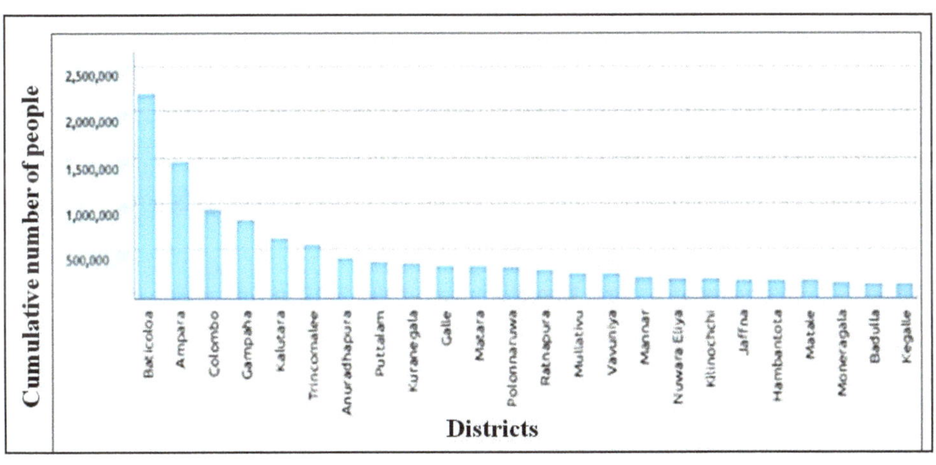

Figure 10.4a: The Cumulative Number of People Affected by Flood during 2002-12.

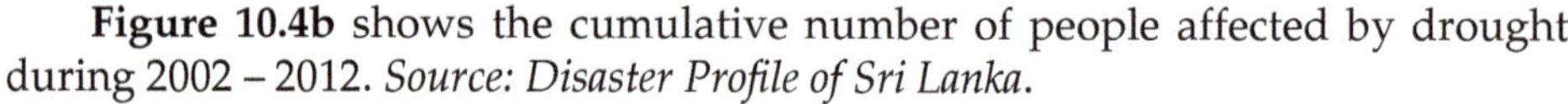

Figure 10.4b shows the cumulative number of people affected by drought during 2002 – 2012. *Source: Disaster Profile of Sri Lanka.*

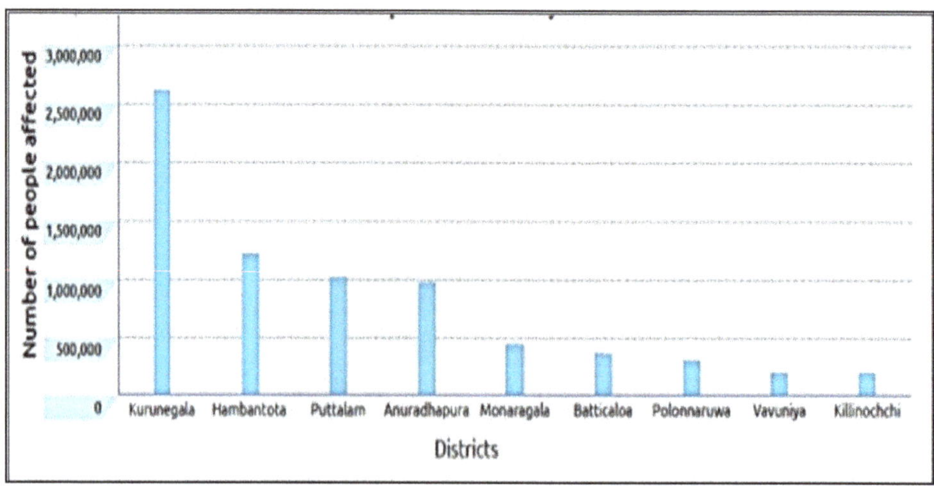

Figure 10.4b: The Cumulative Number of People Affected by Drought during 2002-12.

The losses and damages in Batticaloa, Polonnaruwa, Anuradhapura, and Ampara districts due to flood in 2011. *Source: Disaster Profile of Sri Lanka*

Table 10.1: Losses and Damages in Batticaloa, Polonnaruwa, Anuradhapura, and Ampara District Due to Flood in 2011.

Sector	*Cost of Damage and Losses (LKR million)*
Housing	7,575
Agriculture	15,070
Irrigation	3,000
Road	48,916
Livestock	1,914
Total	**77,475**

Table 10.2 shows the overview of natural disasters from 1980-2010 in Sri Lanka. *Source: Disaster Profile of Sri Lanka*

Table 10.2: Overview of Natural Disasters from 1980-2010 in Sri Lanka

No of Events	**62**
No of people killed	36,982
Average killed per year	1,193
No of people affected	17,457,668
Average affected per year	563.151
Economic damage (US$ × 1,000)	1,674,364
Economic damage per year (US$ × 1,000)	54,012

Section III

Climate Change

Chapter 11

Climate Change and Extreme Events in Afghanistan

N. Sediqi

National Environment Protection Agency,
Darulaman Road, Kabul, Afghanistan
E-mail: eas.div@hotmail.com

ABSTRACT

This research article mainly discusses about the climate change and its impacts in Afghanistan. The main problem is that Afghanistan will be confronted by different new climate hazards based on the current researches. The most likely adverse impacts of climate change in Afghanistan are drought, including associated dynamics of desertification and land degradation. The extreme events and disasters, including flooding, landslides, drought, and extreme heat and freezing weather are the impact of climate change in Afghanistan. This research mainly aims to identify the adverse impact of climate change in Kabul province. The data collection methodology has been conducted through paper reviews as well as an ethnographical interview done in 12 districts of Kabul province. This interview was based on the questions asked from the elders of each districts. The finding from this study shows the percentage of extreme events and climate change impacts in these 12 districts. Moreover, we found that natural resources provide the livelihood basis for up to 80 per cent of the Afghan population. Their sustainable use and management is therefore of essential importance to the well-being of both present and future generations. In addition, the finding from paper reviews showed that the key challenges in impact of climate change are drought, water decrease, reduction of agricultural, and forests, as well as lack of food security. In the conclusion of this study climate change is discussed as a new topic that people have limited knowledge and authorities should increase the public awareness to decrease the climate change impacts.

Keywords: *Mainstream, Land degradation, Hazard, Landslides, Flood, Snowmelt, Deforestation, Vegetation.*

1. Introduction

Afghanistan has an arid and semi-arid continental climate with cold winters and hot summers. The climate varies substantially from one region to another due to dramatic changes in topography. The wet season generally runs from winter through early spring, but the country on the whole is dry falling within the Desert or Desert Steppe climate classification. The snow season averages from October to April in the mountains and varies considerably with elevation.

Recent investigation by the climate change experts shows that Afghanistan will be confronted by a range of new and escalated climatic hazards, the most likely adverse impacts of climate change in Kabul province of Afghanistan are drought, including associated dynamics of desertification and land degradation. The extreme events observed including floods, landslides, drought, and extreme heat and freezing weather are the major impact of climate change in Afghanistan. These events eventually escalated the outbreak of diseases, pollution, social conflicts, and shortage of water resources in most provinces of Afghanistan. These conflicts finally left behind serious confrontations in social, economical, and livelihood of Afghan people.

The main purpose of this research is to identify the impacts and to mitigate the adverse impact of climate change in Kabul Province of Afghanistan. In addition, it aims to search the ways on how to fill the gap for data, proper policy and increasing the public awareness in the current situation of climate change in Afghanistan. The importance of this research is to mainstream climate change and extreme events in Afghanistan into the current sectoral policies and programs of the government of the Islamic Republic of Afghanistan. The main question in this paper that will be researched is about how the impact of climate change affects agriculture, and natural resources and livelihood, besides what are the ways to mitigate climate change risks?

2. Review on Climate Change Issues in Afghanistan

Climate change is a global phenomenon, the effects are local. Physical impacts are determined by geography and micro-level interactions between global warming and existing weather patterns. Human development impacts also vary as changes in climate patterns interact with pre-existing social and economic vulnerabilities (Savage, 2009). In Afghanistan, impacts are likely to be particularly severe due to the arid and semi-arid nature of the country and the extreme poverty within which a large proportion of the Afghan population currently lives. The Second National Communication under the United Nations Framework Convention on Climate Change (UNFCCC) reported that nearly all of Afghanistan's provinces have been affected by at least one natural disaster in the last three decades (SNC, 2017). According to Savage (2009) the principal climate hazard experienced by Afghanistan is drought, which affects all provinces as a result of the high dependence on rain-fed crops. The country experienced repeated droughts in 1963-64, 1966-67, 1970-72, 1998-2006, and 2008-09, which led to significant crop losses. The impacts of the 1998-2006 droughts were particularly severe and resulted in a shortage of food for over 2 million people as well as losses of: (i) 75 per cent of wheat crops, (ii)

85 per cent of rice crops, (iii) 85 per cent of maize crops, (iv) 50 per cent of potato crops and (vi) 60 per cent of overall farm production.

2.1. Recent Climate Hazards Experienced in Kabul Province of Afghanistan

Harvesting Demands on Natural Ecosystems

The reduced productivity of degraded and desertified land reduces the resilience of ecosystems and increases the vulnerability of local communities to the effects of climate change. In Afghanistan desert areas have expanded since 1960 and the incidence of degraded land has increased significantly over the same period (SNC, 2017). Flooding and landslides are common in Afghanistan's mountain and valley regions, particularly during snowmelt in spring and summer. Flooding occurred in 2003 after unusually heavy snowfall, and further flooding in 2005 affected the northern, central and western provinces. Floods and landslides between March and June 2007 claimed 118 lives, affected more than 3,000 families, and destroyed thousands of hectares of agricultural land resulting in 13 provinces being declared disaster areas by the government. Flooding from glacial lake outbursts in the central and northern highlands is also a risk in summer when melt water can break through terminal moraines and cause flash flooding (INC, 2012). Flooding causes rapid erosion of soils, as well as short-term inundation and damage to crops. Furthermore, damage to infrastructure such as roads, bridges and hydro-electric power installations can be extremely difficult and costly to repair and can have substantial consequences for socio-economic development (SNC, 2017). According to the 2006 National Report of the Ministry of Agriculture, Irrigation and Livestock (MAIL), desertification in Afghanistan affects more than 75 per cent of the total land area in northern, western and southern regions where widespread grazing and deforestation have reduced vegetation cover and catalyzed accelerated land degradation.

3. Methodology

In this research study, both interview and recent research reports developed have been reviewed. The main part of this research paper is the latest group study started in 2016 and completed.

3.1. Participants

The target groups for study were 12 districts of Kabul province and according to the similar evidence we have discussed six of them in this paper. These districts were called Charasyab, Paghman, Shakardara, Mirbacha Koat, Surubi, Deh Sabz, Bagrami, Qara Bagh, Guldara, Farza, Kalakan, and Istalif. For conducting this research in Kabul province a group of our colleagues from National Environment Protection Agency of Afghanistan (NEPA), Climate Change division were assigned to collect about different aspects of extreme events and agriculture.

3.2 Instruments

The instruments for conducting this research to assess and collect data we used the ethnographical interviews.

3.3. Procedures

Through this research first of all a professional team was selected by the *Climate Change and Adaptation* (CCA) division of NEPA under my supervision. Then, we prepared a visiting plan for each district and shared with the officials of each districts to inform them about this study. Before traveling to each district the head of district were mandated to inform and invite all districts' elders and had collective interview with their representatives. They shared their experiences, information from recent and past evidences about the disasters as well as their local *Early Warning System* (EWS) in their areas. They talked and responded our queries what we had for them, specifically for each sector and field.

4. Discussion and Findings

The findings regarding climate change and extreme events that we gathered by ethnographical interview from each districts are being discussed as following, **Figure 11.1** shows that in Shakardara district there is a decrease of water surface, karrez, agriculture, forests, and pastures. Besides, the people stated that from 308 karrez 120 of them drought as a result of decrease on water surface. The same changes can be seen in Qarabagh, Guldarah, Meerbacha Koot and Farzah districts, as all these districts are located at the same geographical area and suffering from the similar changing of climate and extreme events **Figure 11.1**.

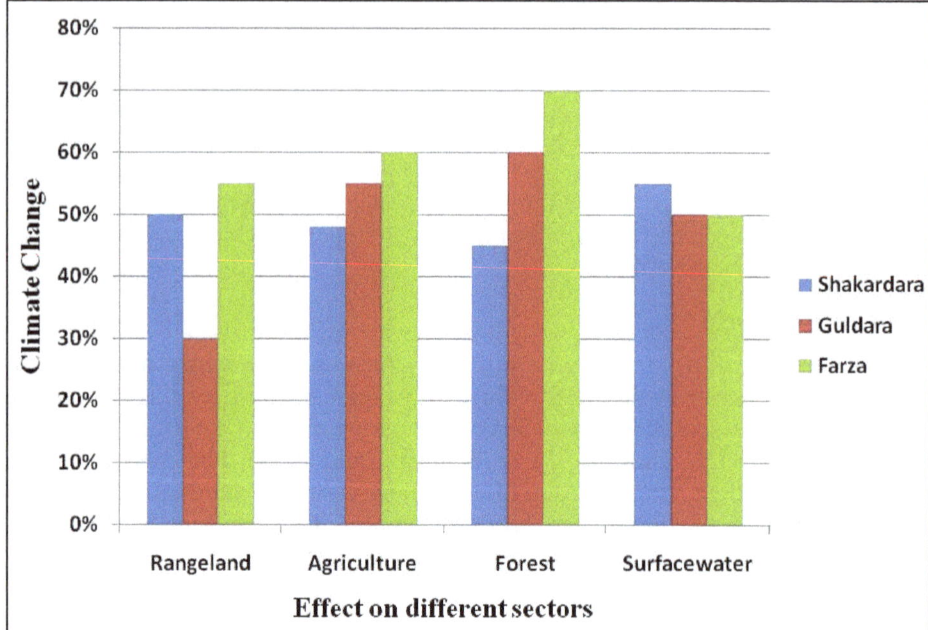

Figure 11.1: Impact of Climate Change and its Effect on Extreme Events in different Sectors, Average Decrease in Percentage.

In addition, in **Figure 11.2** Evidence in Chaharasyab district shows that there is a decrease of 40-50 per cent on the surface of water, karrez, 50-60 per cent in agriculture, forests 45-50 per cent, distinct of fruitful and no fruitful trees, and decrease in rangelands/pastures. Besides, in Paghman district the percentage of underground water is increased up to 80-90 per cent which has paved the ground for rangelands and agriculture.

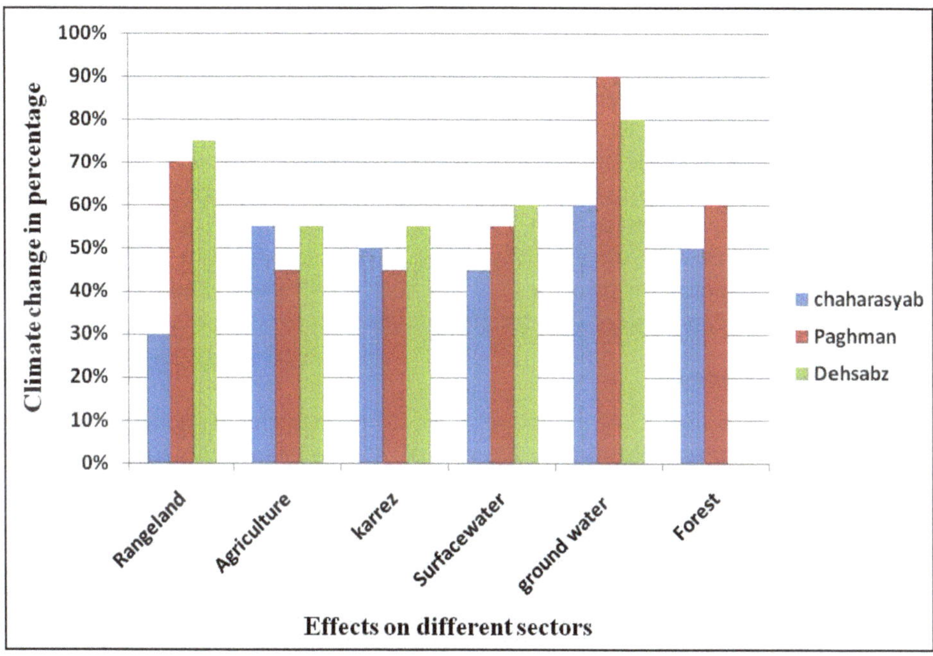

Figure 11.2: Impact of Climate Change and its Effect on Extreme Events in different Sectors; Average Decrease in Percentage.

On the other hand, in Bagrami district approximately around 90 per cent decrease has been seen on the ground water. There is also a decrease in agriculture and surface water compared to the past. Another important change in this district is building of residential houses that destroyed rangelands.

Furthermore, the key challenges in impact of climate change are drought, water decrease, reduction of agricultural, and forests, as well as lack of food security. In addition, increase of diseases, lack of adaptation, and lack of resistance, increase in level of vulnerability, and a harmful effect on farming which reduce the financial situation of farmers. Finally, it increases the number of natural disaster.

Moreover, natural resources provide the livelihood basis for up to 80 per cent of the Afghan population. Their sustainable use and management is therefore of essential importance to the well-being of both present and future generations. At the same time, Afghanistan's natural resources are being degraded due to the extreme events and climate change, internal displacement, high rates of population growth, low levels of education, and poverty, result in a prioritization of survival over the longer-term sustainability of natural resource use and management. This has devastating consequences, particularly in an environment where the balance between precipitation and primary production is very finely balanced as the effects of natural disasters, such as drought and flooding, are magnified many times when the ability of the natural resource base and associated livelihoods to recuperate have already been weakened.

4. Conclusions

Overall, as the outcome of this research we found that in the Afghanistan there is limited knowledge of climate change and its effects. It is often referred to as being a new topic. However, to be integrated into national policies and strategies a more comprehensive understanding of climate change and options for addressing it are necessary for policy makers and across ministries. Workshops will raise the awareness of policy makers and other government ministry staff on climate change and extreme events. As well as, building capacity to mainstream climate change into national development priorities and should be increased. Finally, raising the public awareness can prevent the climate change risks up to some extent.

Acknowledgement

I would like to appreciate the cooperation of my department and head of NEPA (National Environment Protection Agency of Afghanistan) who supported and provided the opportunity to travel and gather information I needed.

References

1. INC, 2012. Afghanistan Initial National Communication (INC) Report. To the United Nations Framework Convention on Climate Change (UNFCCC), National Environmental Protection Agency of the Islamic Republic of Afghanistan.

   https://unfccc.int/resource/docs/natc/afgnc1.pdf

2. SNC, 2017. Second National Communication (SNC) Report. To the United Nations Framework Convention on Climate Change (UNFCCC), National Environmental Protection Agency of the Islamic Republic of Afghanistan.

   https://postconflict.unep.ch/publications/Afghanistan/Second_National_Communication_Report2018.pdf

3. Savage M., Dougherty B., Hamza M., Butterfield R., Bharwani S. 2009. Socio economic impacts of climate change in Afghanistan. A Report to the Department for International Development by Stockholm Environment Institute, Kräftriket, Stockholm, Sweden.

   https://www.weadapt.org/sites/weadapt.org/files/legacy-new/placemarks/files/5345354491559sei-dfid-afghanistan-report-1-.pdf

Chapter 12

Climate Change and its Impact on the Economic Sectors in the Rural Areas of Karnataka, India: The Need for Multi-Pronged Approach

B.C. Prabhakar[1]* and K.N. Radhika[2]

[1]Department of Geology, Bangalore University, Bangalore – 560 056, Karnataka, India
[2]Department of Civil Engineering, East West Institute of Technology, Bangalore – 560 091, Karnataka, India
****E-mail: bcprabhakar@rediffmail.com***

ABSTRACT

With the advent of global warming and climate change, the entire world is affected and much of the impact is felt by developing countries as they basically depend on agriculture for the livelihood and economy. Climate change is affecting India in many sectors like agriculture, water resources, ecosystems, coastal zones etc. Karnataka, the 7th largest State in India with a rural population of about 37.5 million accounts to 61.43 per cent of the total State's population is the second least rain fed region in India. As in many other states, recurring droughts and poor crop productivity is prevalent in Karnataka and incidences of farmers committing suicide is also high. This is signaling towards one major phenomenon – the climate change. In the past 16 years (2000-2016), the state has faced drought for 13 years. In the year 2016, the State had declared 139 out of 176 talukas as drought hit and sought (from Central Govt.) 4,702 crores Indian Rupees for the loss of crops due to severe drought. The drought conditions have also severely affected agricultural activities in 123 taluks during 2011, 157 taluks during 2012, 125 taluks during 2013, 35 taluks during 2014 and 136 taluks during 2015. The impact of drought on the economy was evident at both macro and micro level. Total estimated loss during 2016 drought in Karnataka was 14,471 crores Indian rupees. The droughts had potential implications on government policies, via their impact on the budgetary balance. Of all, ensuring water needs during drought period was the mostcrucial issue. To develop adaptation strategies for rural communities to face droughts, it is necessary to cope with current climate-related risks, besides taking clues from the age-old coping strategies that the farmers and rural households would have developed. Strategizing the management plans for vulnerable communities become very important in this context.

Keywords: *Climate change, Drought, Agriculture, Rural sector, India, Economy.*

1. Introduction

Climate changes have been occurring ever since the atmosphere was created on earth about 3.5 billion years ago. The early climate changes were influenced only by natural phenomena, prior to the origin of man. These climate changes were responsible for cold and warm conditions which prevailed for long periods and are recorded in the geologic formations. But in the modern era, the natural phenomena along with human contributions have emerged as a strong contributing factor in influencing the climate change, especially in warming up the globe. The human impact on climate is so alarming that in the last 100 years, the mean global surface temperature has risen by 0.4-0.8°C (IPCC, 2001). The IPCC report also notes that the global mean temperature may increase anywhere between 1.4 and 5.8°C by 2100. This change could transform the ecosystems with catastrophic disruptions on livelihoods, economic activity and living conditions. Going by the present trend of population increase, urbanization and the consequent industrial expansion, it is highly unlikely that the global mean warming could be stabilized below 1.5 to 2 °C (IPCC, 2001). Because, there has been a constant raise in the green house gas (GHGs) addition to the atmosphere largely due to automobiles and industrial emissions.

Globally, half of the world's population *i.e.*, approximately 3.3 billion people live in rural areas and 90 per cent of those people reside in developing countries. Thus, with the advent of global warming and climate change, much of the impact is felt by developing countries as they basically depend on agriculture for the livelihood and economy. As an example, nearly 40 per cent of the GDP of India comes from agrarian sector. The two major global climate change impacts in rural areas are (1) effect on the infrastructure and (2) the impact on agriculture and other ecosystems. The effect on rural infrastructure originates due to natural calamities like floods, storms, landslides *etc.*, whereas, the impact on agriculture and other allied activities like keeping livestock, forestry, fishery, wild life conservation *etc.* would face serious constrains due to failure of rains and the consequent droughts.

Across Asia one billion people could face water shortage leading to drought and land degradation by 2050 (Christensen *et al.*, 2007; Cruz *et al.*, 2007). India being largely an agriculture-dominant country is facing recurring drought conditions leading to extreme hardships to people in rural areas, besides impacting the surrounding in innumerable ways. India being a welfare state and with almost 22 per cent of the population belongs to below poverty line, obviously demand strategic planning and management to meet the impacts caused by climate change. The global assessment report (UNISDR, 2015) produced by the UN office for Disaster Risk reduction reports that India's average annual economic loss due to disasters is estimated to be US $ 9.8 billion and has urged all the Asian countries to treat this as a wakeup call and make adequate investment in Disaster Risk Reduction (DRR) or it will hinder the development process of the countries.

In this paper, the authors present how the rural areas face huge losses due to the drought conditions in Karnataka, a Southern State in India, and their impact on national economy and various ways through which we can mitigate the effects.

2. Vulnerability of Rural Sectors for Drought in INDIA

India, especially the rural India is vulnerable to extreme climatic impacts since the infrastructure built to face them is poor and economic strength of individual families are dismal. About120 million ha space in 185 districts covering 1173 talukas are repeatedly subjected for droughts (**Figure 12.1**). India has faced 28 drought years during 20th Century which speaks to the severity of problem. It may be worth recalling that there were 25 major famines across India, which killed 30-40 million people in later half of the 19th century. The first Bengal famine (1770) wiped out nearly one third of the population and the next devastating Bengal famine (1943–44) affected 3-4 million people.

India has a total rural population of 68.84 per cent (as per 2011 census) and agriculture is contributing significantly for the country's gross domestic product (GDP). But, climate change has a profound impact on the Indian economy as droughts are prevailing periodically along with secondary problems like water scarcity, reduced crop yield, distress to people *etc.* The best example for this is the drought in the year 2015 which affected many of the leading crop producing States like Punjab, Uttar Pradesh, and Maharashtra *etc.* with an overall reduction of crop yield by 5 per cent. An estimate indicates that the 2015-16 drought cost national economy about 6,50,000 Crores Indian Rupees. About 330 million people were affected by this drought. It created a serious challenge for the drought affected areas in 10 states, the worst hit were Maharashtra and Karnataka.

Projected impacts of climate change on Indian agriculture envisage a decrease in cereal productivity by 10-40 per cent by 2100. Greater loss is also likely in Rabi crops. Every 1° C increase in temperature reduces wheat production by 4-5 million tones. Increased river and sea water temperature affect fish breeding, migration and harvests, besides affecting milk production.

3. Recurring Drought in Karnataka

Karnataka's agriculture is mainly rain fed with almost 68 per cent of its farmland without irrigation. 54 per cent of Karnataka's geographical area is drought prone. 88 of the State's 176 Taluks and 18 of its 30 districts are drought-prone, impacting crop productivity, forest biodiversity, hydrological cycle and health of people.

Karnataka, being the 7th largest State in India with a rural population of about 37.5 million accounting to 61.43 per cent of the total State's population is the second least rainfall receiving region in India. Geographically, broadly the State is divided into 4 regions *viz.*, South interior, North interior, Malnad and Coastal Regions. The annual rainfall varies both spatially and temporally with highest rainfall of about 4747 mm in coastal region followed by Malnad region with 3500 mm to as low as

477mm in south interior Karnataka. In the past 16 years (2000 – 2016), the State has faced droughts for 13 years (Report of the KSNDMC, 2017). In the year 2016, as per the parameters prescribed in the drought manual (GOI, 2009), 139 out of 176 Taluks were declared as drought affected. The parameters measured to consider drought included 3 aspects *viz.* (1) meteorologically long term average of 25 per cent or less rainfall as normal; 26-50 per cent as moderate and more than 50 per cent as severe drought; (2) deficiencies in surface and sub-surface water supplies and (3) 4 consecutive weeks of poor weekly rainfall of nearly 50 mm with poor sowing. With all the 3 factors, mentioned above affecting the State, the Govt. of Karnataka sought (from Central Govt.) 4,702 Crores Indian Rupees for the loss of crops, and to initiate remedial measures in the rural sectors. This drought was recorded worst in past 40 years. The areas hit by drought and deficiency of rainfall are shown in **Figure 12.1** and **Table 12.1**. Almost the entire State faced the drought impact in 2016. Prior to 2016, the State had experienced severe drought conditions consecutively during 2011 to 2015 (**Figure 12.2**). Many parts of the state, including Malnad and Coastal districts received deficit rainfall due to monsoon failure during successive years. 22 taluks of 7 districts were subjected to droughts successively during the last 5 years. The drought conditions have severely affected agriculture in 123 taluks during 2011, 157 taluks during 2012, 125 taluks during 2013, 35 taluks during 2014 and 136 taluks during 2015. As in many other states, recurring droughts and loss of crops, incidences of farmers committing suicide were also high.

Table 12.1: Cumulative RF Pattern from June to October 2016 in Karnataka Revealing Severe Drought Conditions

Sl.No.	*Karnataka State Zone*	*Cumulative RF Pattern June to October 2016*		
		Normal (mm)	*Actual (mm)*	*per cent DEP*
1	South Interior Karnataka	505	336	-33
2	North Interior Karnataka	605	485	-20
3	Malnad	1665	1092	-34
4	Coastal	3206	2476	-23
Total		**975**	**718**	**-26**

Source: KSNDMC.

4. Impact of Drought on Rural Economy

The impact of drought on the economy was evident at both macro and micro level. The overall impact on the state is perceived as a macro level impact where as impacts at village and household level as micro-level. The impact was also either direct or indirect which varied in nature and intensity. Social structure (class and caste), village and household resource endowments also mattered in the degree of impact. Direct impacts were evident on physical, social, economic and environmental aspects. To consider the physical aspects, agriculture was the first to be affected by drought. Huge crop losses in 30.82 lakh hectares in 139 taluks of 25 districts were reported due to rainfall deficit. Within the agricultural sector, marginal (nearly 10 acres) and small farms (nearly 5 acres) were more vulnerable, because of their

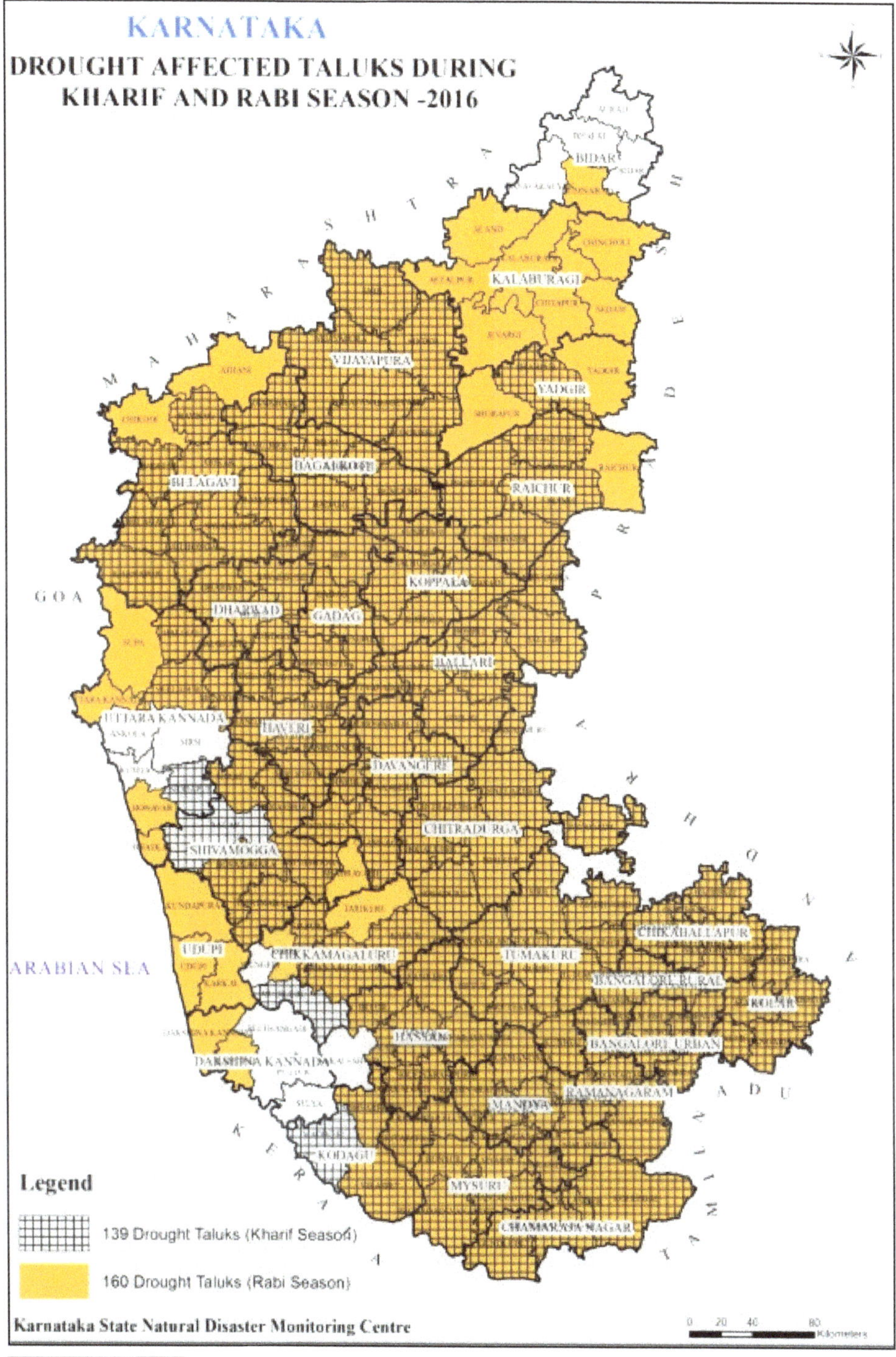

Figure 12.1: Map of Karnataka State Showing the Incidences of Droughts in 2016 (after KSNDMC, 2017).

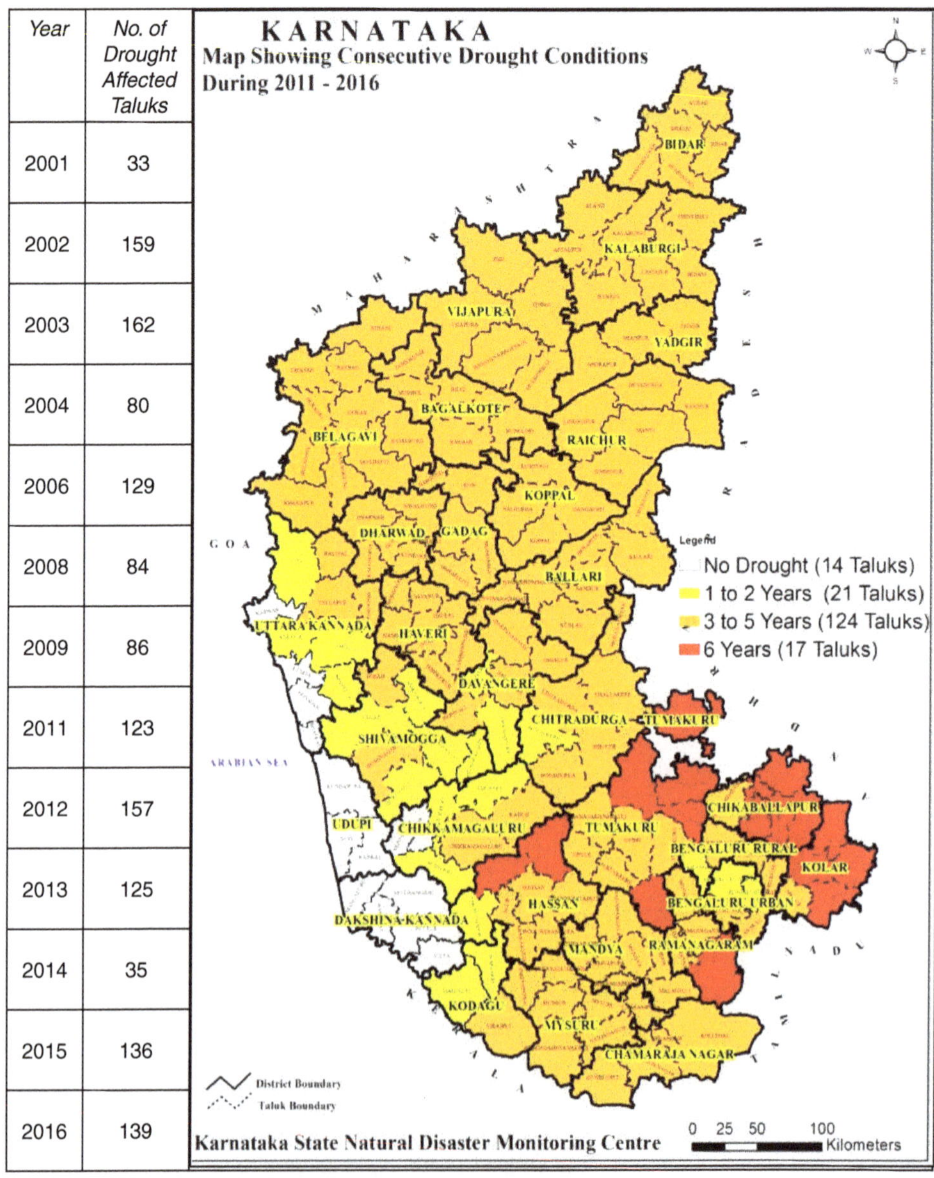

Year	No. of Drought Affected Taluks
2001	33
2002	159
2003	162
2004	80
2006	129
2008	84
2009	86
2011	123
2012	157
2013	125
2014	35
2015	136
2016	139

Figure 12.2: Map Showing Incidences of Drought in Karnataka from 2011 to 2016 (*Source*: KSNDMC Report, 2017).

dependence on rainfed agriculture and related activities. Loss of assets in the form of livestock and productive capital damage were also the other forms of physical losses besides impacting agro-based industries in rural sector. Domestic availability of water was greatly restricted which in turn had implications on health of households.

There were serious impacts on social aspects as well. Law and order services were put under strain by a rise in crime. Temporary unemployment, migration, increased destitution, spurt in child marriages and food insecurity were rampant in drought-hit areas. As per the statistics of Agriculture Department 1,002 farmers have ended their lives during April 2015 to January 2016, the severest drought period in recent times. With this, Karnataka broke all the previous records, as far as farmers' suicide is concerned and drought was one of the potential reasons to push the hapless farmers to take the extreme step. **Table 12.3** shows farm suicides in India from 2013 to 2016, the period India witnessed some of the worst droughts.

Table 12.3: Farm Suicides in India

Year	*Total Farm Suicides (Farmers and Agricultural Laborers)*	*Source*
2013	11772	NCRB*, ASDI** annual reports
2014	12360	
2015	12602	
2016	6667 from just 5 states – Madhya Pradesh, Maharashtra, Karnataka and Telangana + Tamil Nadu	

* NCRB- National Crime Record Bureau; ** ADSI – Accidental Deaths and Suicides in India.

Economic impacts both at individual/family level and at State level were severe. As mentioned earlier, total estimated loss during 2016 drought in Karnataka was 14,471 Crores Indian Rupees. The droughts had potential implications for government policies, via their impact on the budgetary balance. It (drought) reduced tax revenue, through a decline in income, employment and exports. On the expenditure side, increased spending on drought relief, social welfare, water supplies, subsidies, *etc.* became inevitable. 694.12 Crores Indian Rupees of crop insurance due to loss in the Kharif season (covers < 50 per cent of affected farmers) alone was to be paid to the farmers.

In the environmental front too there were telling effects. Farming community was affected by destruction of ecosystems, on which agriculture and other related activities are closely linked. Because they depend on natural surroundings for many needs – grasslands, forest for foliage, water, manure *etc.*

The State of Karnataka, especially the northern region witnessed indirect impacts of droughts in many ways. Regional inequality leading to a cry for a separate State had its roots in the economic disparity, which in turn was due to recurring droughts. The political groups in the northern parts of the State felt that a separate Statehood for themselves could enable to create a better living conditions. As their purchasing power declined, unemployment increased, reducing the availability of debts. Exodus of rural people to urban places has seriously affected theinadequately built infrastructure in cities like Bangalore. At the same time, the migrated vulnerable group was forced to work at lower wages or live in near-hunger conditions. The intensity varied according to their economic strength, that is, the ability of households to cope with drought. Studies revealed that the impact of drought were more on women, children and households belonging to Scheduled Castes and Tribes, as most of these groups were already in a disadvantaged position in society.

5. Climate Change Forecast for Karnataka

Projections (**Figure 12.3**) of climate change based on an ensemble mean of 18 Global Climate Models (GCMs), have been attempted by Centre for Sustainable Technologies (CST) in Indian Institute of Science by considering the temperature variations from 1861 and projecting up to 2099 (CST, 2014). Representative concentration pathways (RCP), which constitute new scenarios of the emissions of greenhouse gases (GHGs), represent different levels of radiative forcing leading to different levels of warming (2.6, 4.5, 6.0, and 8.5 Watts/m^2).

From this model, the temperature trends over the last century suggest that Karnataka has warmed by about 0.4°C. In the short term (2030s), the temperature is projected to increase by about 1.7 °C relative to that in pre-industrial period (1880s). The long term *i.e.* in 2080s, the mean temperature increase for Karnataka could be as high as 5.0 °C by the end of the century under the high-emissions scenario (RCP 8.5) relative to that in the pre-industrial period. Most parts of Karnataka could experience 1.5 - 2°C warming by as early as the 2030s, relative to that in the pre-industrial period (*i.e.*,1880s), under the high-emissions scenarios. The rainfall trends over the last century indicate an overall decrease in annual rainfall in Karnataka by 10 per cent. The decrease is more than 10 per cent in coastal and south interior districts like Chikmagalur, Dakshina Kannada, Kodagu, Shimoga, Uttara Kannada, and Udupi. Rainfall has also declined significantly in northern districts such as Bidar, Bijapur, Gulbarga, and Yadgir.

Thus, going by the past records, present climatic variations and projected future trends the threat of drought looms large over most part of Karnataka. Because there are no permanent measures established to handle the situation, most drought relief programs are undertaken as contingent measures, but not as long term strategies. Large section of the rural population still believe that it is an inevitable situation with which they have to live in. But it is the responsibility of the governments and civil society to chalk out measures and implement to mitigate the impacts of droughts.

6. Strategies to Cope with Current Climate Risk

To develop adaptation strategies for rural communities to face droughts, it is necessary to cope with current climate related risks, besides taking clues from the age old coping strategies that the farmers and rural households would have developed. Strategizing the management plans for vulnerable communities become very important in this context. These strategies can be (1) providing skill-based training, (2) job creation for the people of affected areas, (3) strengthening the National Literacy Policy, especially to promote literacy among women, (4) policies aimed at benefitting the deprived population (in terms of housing, food, subsidies, *etc.*), (5) strengthening of other sources of income besides agriculture, (6) reducing population growth by family planning and making people aware of the benefits of small families, and (7) dairy and live stock activities that have greater market linkages and government support. Time tested methods like shift in cropping

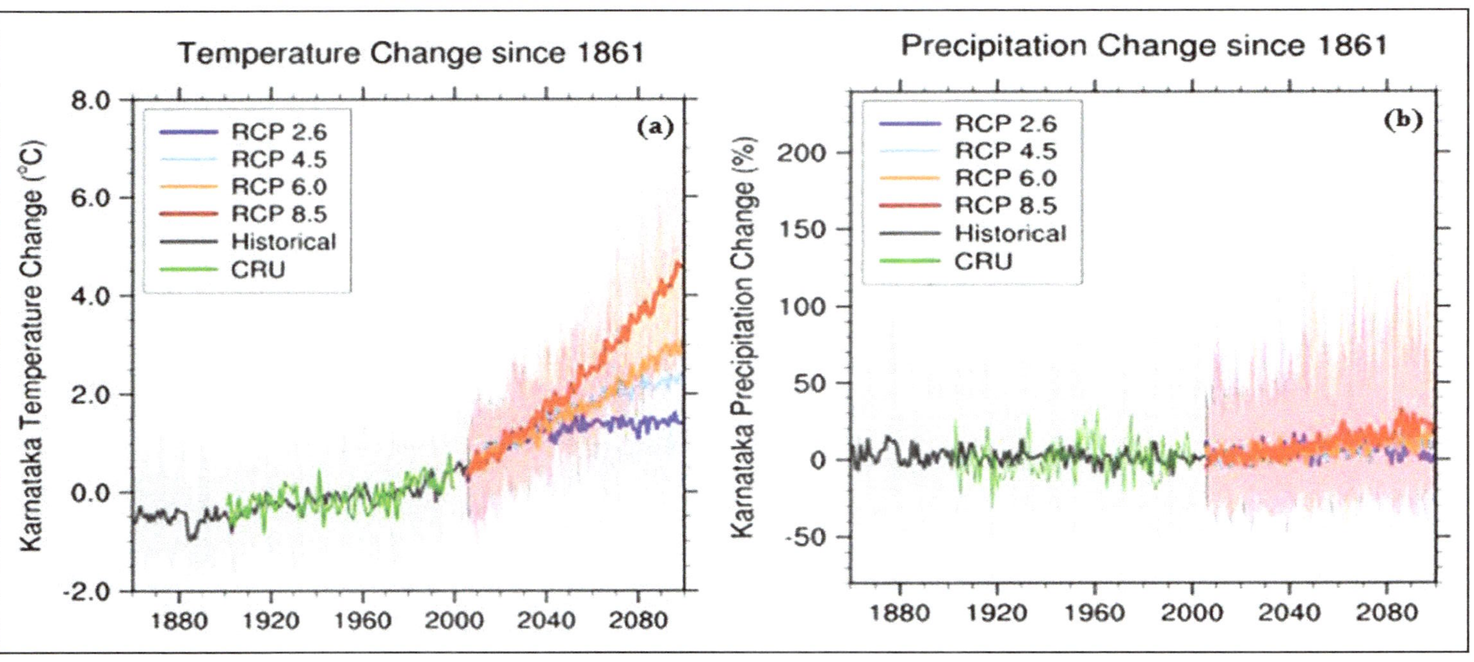

Figure 12.3: Projections of Climate Change Based on an Ensemble Mean of 18 Global Climate Models (*Source*: CST, 2014).

patterns, crop diversification, mixed cropping and changes in the date of sowing should also be pursued strongly by farming community and the government should support through wide spread propaganda in rural areas.

The project design of rural development programmes needs a radical change and should encompass technology based approaches for faster reaching out and implementation of various remedial measures. There should be a development based approach aimed at reducing poverty and vulnerability. Ecosystem based approach is also critical which includes promoting biodiversity and ecosystem services, as they are critical in conserving agrarian environment.

Forest being the nerve controller of climate changes, it is essential to conserve the forest cover of the State.Forests in Karnataka account for 19 per cent of the geographic area (compared to a national average of 21 per cent) of which about 56 per cent is dense forest and about 44 per cent open and scrubby forests. Even as Karnataka is implementing afforestation programmes at around 40,000 ha annually over the past decade, forest area in Karnataka has declined by about 2 per cent during this period and in addition large parts of dense forests have been converted to open and degraded forests. Welfare policies of the State though aiming at helping the rural poor (through allotting forest land for agriculture), are doing a big damage to forests both on macro and microscale, which in turn are contributing to climate change at regional level. **Table 12.4** shows the vulnerability of forests in Karnataka and stresses on the urgent need to take stern measure to check the deterioration of forest cover.

Table 12.4: Forest Grids in Low, Medium, High, and Very High Vulnerability Classes, by Forest Type (CST Report, 2014)

Regions	*Inherent Vulnerability (Per cent of grids)*			
	Very High	*High*	*Medium*	*Low*
Evergreen	2.85	8.99	25.44	62.72
Semi-evergreen	0.00	10.69	35.85	53.46
Most deciduous	1.92	11.82	51.26	35.01
Dry deciduous	10.64	36.78	48.63	3.95
Plantations	40.35	24.50	22.24	12.92

After, IISC, 2014.

Of all, ensuring water needs during drought period is the most crucial issue. Interlinking rivers; diversion of rivers; restoring/rejuvenating lakes; developing feeder channels to link lakes; restoring river routes; enhancing groundwater potential and developing deep community wells have been the strategies that could help fight water scarcity. But when the rains fail for long periods, we should have developed plans and infrastructure to tap water from oceans through desalination. Currently, more than 7500 desalination plants are operating around the world accounting for a worldwide water production of 65.2 million m^3/year (0.6 per cent of global water supply). Karnataka with about 300 Km coastal line can meet all the water requirements through desalination, if we can establish elaborate infrastructure like in Israel. Though the initial budget of desalination projects looks colossal, the advantages are many. The major component of expenditure is for the energy to desalinateand transporting to the needy areas. Karnataka, situated in tropical zone, enjoys abundant sun shine which can be harnessed by establishing solar panels in feasible areas to tap solar energy to use it for desalination. Though laying pipes to transport desalinated water is a challenging task, it will be greatly rewarded when drought hit areas are assured of at least drinking water. What is required is prioritizing the National/State budgets to achieve such challenging tasks. Such ideas and implementation could yield far reaching benefits on humans, live-stock, food production, social security and environment especially in rural areas which bear most of the brunt of climate change and consequent droughts. It also lessens frequent spending on relief measures, which are but temporary and hardly help in the alleviation of poverty and providing conditions for better living.

7. Conclusions

Karnataka, like many States of India, is vulnerable for extreme climatic changes and drought is one such condition which strikes often. Drought will impact all sectors and all communities at various levels. Hence, considerations of climate change need to be incorporated into all developmental programmes. Scientific rationale should be in forefront in the management of extreme natural events, than political decisions. Prioritization of budget for long term/permanent plans to mitigate droughts should be a priority issue. Conventional and technological adaptation strategies are the need of the hour which can be complimentary and effective for agriculture, water, forests, rural development, and health during the combat against droughts. Resource pooling, economic restructuring and establishing effective management system could be some of the additionallong term strategies to minimize the impact of climate change and consequent droughts.

References

1. CST Report, 2014. Transitioning towards climate resilient development in Karnataka. Publication Centre for Sustainable Technologies, IISC Bangalore. pp.3-14. http://www.greengrowthknowledge.org
2. Christensen, J.H., Hewitson, B., Busuoic, A., Chen, A., Gao, X., Held, I. and Whetton, P. 2007. Regional climate projections. In S. Solomon, D. Qin, M. Manning, Z. Chen, M. Marquis, K. B. Averyt, M.Tignor, and H. L. Miller (Eds.), Climate change 2007: The physical science basis. Contribution of working group I to the Fourth Assessment Report of the Intergovernmental Panel on Climate change (pp.847-940). Cambridge/New York: Cambridge University Press.
3. Cruz, R.V., Harasawa, H., Lal. M., Wu, S., Anokhin, Y., Punsalmaa, B. and C.E. Hanson (Eds.). Climate change 2007. Impacts, adaptation and vulnerability. Contribution of working group II to the fourth assessment report of the Intergovernmental Panel on Climate Change (pp. 469 – 506). Cambridge University Press, Cambridge, United Kingdom.

4. GOI 2009. Manuel of drought management. Published by, Department of Agriculture and Cooperation, Government of India, New Delhi. Pp.149. http://nidm.gov.in

5. IPCC 2001. Climate Change 2001: The Scientific Basis. Contribution of Working Group I to the Third Assessment Report of the Intergovernmental Panel on Climate Change [Houghton, J.T., Y. Ding, D.J. Griggs, M. Noguer, P.J. van der Linden, X. Dai, K. Maskell, and C.A. Johnson (eds.)]. Cambridge University Press, Cambridge, United Kingdom and New York, NY, USA, 881pp.

6. KSNDMC, 2017. Drought vulnerability assessment in Karnataka (A composite index using climate, soil, crop cover and livelihood components). Published by Karnataka State Natural Disaster Monitoring centre, Bangalore, pp. 1-97.

7. UNISDR 2015. Making Development Sustainable: The future of disaster risk management. Global Assessment Report on Disaster Risk Reduction. Geneva, Switzerland: United Nations Office for Disaster Risk Reduction (UNISDR). ISBN 978-92-1-132042-8 pp.266.

Chapter 13

Working Out Ways to Find Commercially Viable Measures to Develop Water Resources in the Face of Climate Change and Recurring Droughts in India: An Overview

K.N. Radhika[1] *and B.C. Prabhakar*[2]

[1]*Department of Civil Engineering., East West Institute of Technology, Bengaluru, Karnataka, India*

[2]*Department of Geology, Bangalore University, Bangalore, Karnataka, India*

E-mail: knradhika13@gmail.com

ABSTRACT

Climate change is one of the most widely discussed issues in the present context and its role is more and more realized in understanding and analyzing several vagaries of earth, the chief being scarcity of water leading to drought situation in several parts of the country. Huge sums, which otherwise could have been spent for developmental programs, are diverted to provide bare minimum water needs in the severely affected areas, especially in rural sectors. There is a growing scarcity of water and the demand by 2050 is projected to be 1447 billion cubic metres while the available water resources stand at 1250 billion cubic metres. This imbalance is projected to grow exponentially due to the unprecendented growth in population and rapid urbanization. Besides, the outlook towards the development of rural sectors has not seen radical change since independence and the villages continue to undergo the ordeal of social and economic debacles in the event of the recurring droughts. Hence, there should be effective integrated programs to alleviate the rural populace from the impacts of droughts through political and policy reforms and adopting technological know-how to develop various sources of water. Innovative approaches like desalination, interlinking of rivers, diversion of rivers, cloud ionization, developing deep aquifers, recycling of wastewater etc. should contribute to a better and sustainable management and on the other hand, conventional methods like harnessing of surface water and artificial recharge also should be practiced rigorously. This paper dwells upon the aforesaid issues in the context of the country at large and Karnataka State in particular.

Keywords: *Water scarcity, Global warming, Climate change, Droughts, India, Karnataka, Innovative approaches.*

1. Introduction

Climate change is the net result of many processes taking place on the earth ever since its origin. However, in the recent times, there is a growing concern about the anthropogenic contributions to the climate change outcomes. The industrialization that started during the late 17th century has heralded the process of climate change by emissions of greenhouse gases (GHGs) to the atmosphere. Since then, there has been a manifold increase in the industrial activities and the consequent GHG additions, severely affecting the hydrological cycle and biodiversity, which in turn are responsible for causing extreme hazards like droughts, floods, heat waves *etc.* Oldenborgh, *et al.* (2018), working on the extreme heat wave incidences of India in 2016, forecast their recurrence due to the increasing trend of global warming. Gupta *et al.* (2011), discussing the drought-related disasters of India, state that 13 States in India are experiencing recurring droughts due to climate change, natural resource degradation, improper agricultural practices and poor water management. Thus, failure or erratic rainfall and the critical shortage of water havepushed several parts of India to severe droughts in the recent years, which are the major challenges for government to handle.

Drought is a creeping phenomenon which could amplify into a hazardous one. Approximately 85 per cent of the natural hazards are related to extreme meteorological events (Obasi, 1994) and drought perhaps causes the worst long-term impacts. With the onset of drought, the increased temperature and decreased humidity will progressively aggravate its severity, thus creating a stress on the water resources of the region. With rise in the frequency, intensity and duration of droughts, there would be profound impact on three life sustaining sectors *viz.*, water, food and energy.

When the water and its cycle occupy the center stage of any discussion on climate change and the resulting impacts, it is necessary to appraise the water scenario of the earth. The total volume of water on earth is estimated at 1.386 billion km^3 (Eakins and Sharman, 2010) of which only 2.5 per cent *i.e.* 34.65 million km^3 is fresh water, of which 0.3 per cent (0.10395 million km^3) is present on the surface of the earth and the rest is trapped in the glaciers and ice bergs. In natural purification system of water through the hydrological cycle, around 505,000 km^3 is evaporated from the oceans, and 72,000 km^{3}from land surfaces with a total of around 577,000 km^3/year. Of this, approximately 458,000 km^3/year (80 per cent) falls back into the oceans and only 20 per cent (119,000 km^3/year) falls onto the land. Of the 20 per cent freshwater falling on land as precipitation, most of it is transpired back into the atmosphere almost immediately, leaving only 8 per cent of the total volume on the ground. Much of this forms groundwater that may become inaccessible. Unfortunately rampant pollution of surface and subsurface waters is also posing a threat on the ability of the natural hydrological cycle, in affecting the potential of water resources.

Although global attention has focused primarily on water quantity, efficient water-use and allocation issues, the deterioration of water quality on the other hand due to poor wastewater management, inefficient management of solid waste and unscientific agricultural practices has created serious problems in many parts of the world, worsening the water crisis. Thus, water pollution is a yet another global challenge, undermining economic growth as well as the physical and environmental health of billions of people. Globally, 80 per cent of municipal wastewater is discharged into water bodies untreated. Similarly, industries dump millions of tonnes of heavy metals, solvents, toxic sludge and other wastes into water bodies each year (WWAP, 2017). Agriculture, which accounts for 70 per cent of water abstractions worldwide, also plays a major role in water pollution. Farms discharge large quantities of agrochemicals, organic matter, drug residues, sediments and saline drainage into water bodies. The resultant water pollution poses risks to aquatic ecosystems, human health and productive activities (UNEP, 2016).

Currently, the annual global fresh water demand is 640 billion liters and about 1.2 billion people lack access to clean drinking water. According to the World Water Development Agency classification, a country can be categorized as 'water stressed' when per capita water availability is less than 1700 m^3/year, and 'water scarce' if the per capita supply is less than 1000 m^3. Thus, by 2025, globally about 2.5 billion and 1.8 billion population will be under water-stressed and water-scarce positions, respectively (WWDR, 2015). This acute water scarcity may lead to several conflicts.

2. Global Water Crisis and the Indian Scenario

A recent alarm, has indicated that nine mega cities in the world namely, Cape Town in South Africa, Karachi in Pakistan, Sao Paulo in Brazil, Istanbul in Turkey, Sana in Yemen, Kabul in Afghanistan, Mexico City in Mexico, Buenos Aires in Argentina, Nairobi in Kenya, Beijing in China and Bengaluru in India are heading towards big water crises and pose great developmental challenges for the concerned country. Though copious water potential exists in some countries like Brazil, Russia, United States, Canada, China *etc.*, their uneven distribution also makes the water management task challenging.

Climate change in developing asian countries has been affecting many sectors like agriculture, water resources, ecosystems, coastal zones *etc.* According to a report by the Asian Development Bank (ADB) and the Potsdam Institute for Climate Impact Research (ADB-PIK, 2018), Southern India may witness a decline in rice yields by 5 per cent upto 2030s, 14.5 per cent by 2050s and 17 per cent by the 2080s and the report also project yield reduction of wheat by 8 per cent due to climate change and consequent droughts in India. Thus, woes due to water scarcity in India are looming large. An appraisal of this situation is linked to four main factors, *viz.* (1) the ever inflating population, (2) rapid industrialization and urbanization, (3) the spatial incongruity of population and water resources, and (4) the deteriorating water quality. Studies project India to surpass China by 2022 with the current growth rate of population (World Bank, 2010). By 2050, the country may have to feed an extra population of 500 million. The National Commission for Integrated Water Resources Development of India states that there is a 30 per cent decrease in per capita water availability over the years. In 1951, the per capita water availability was 5200 cubic meters, whereas now it is 1545 cubic meters, and is projected to decrease to 1220 cubic meters by 2050 (**Figure 13.1**). The converse relationship between the decadal increase in population and decrease in precipitate water availability is represented in **Figure 13.2**. Thus, due to the increase in population and decrease in water availability, by 2050, the water demand in India is likely to increase from the current 962 billion cubic meters to 1447 billion cubic meters. Industrialization in India has no doubt strengthened the economy of the country but is gradually inducing the climate change effects. Urbanization is also rampant and currently around 50 per cent of people live in the cities in India, further stressing on the water resources of the country. Exodus of people from rural to urban places depletes water resources, especially ground water, which may lead to urban droughts. Central Ground Water Board (CGWB, 2017) has assessed the net groundwater availability of India by dividing the country into 6584 units (blocks or mandals or talukas) and out of this, 1034 units falling in 16 states and 2 union territories are over-exploited; 253 units are in critical stage; while 681 are in the semi-critical stage, with the net annual groundwater availability to be 62 per cent (**Figure 13.3**). The study by Amarasinghe, *et al.* (2007) also projects the decline of groundwater availability from the current 62 to 26 and 16 per cent by 2025 and 2050, respectively. The groundwater availability of Karnataka is 66 per cent with about 20 per cent of the state falling into the semi-critical and critical categories and 20 per cent of the area is over-exploited (CGWB, 2017).

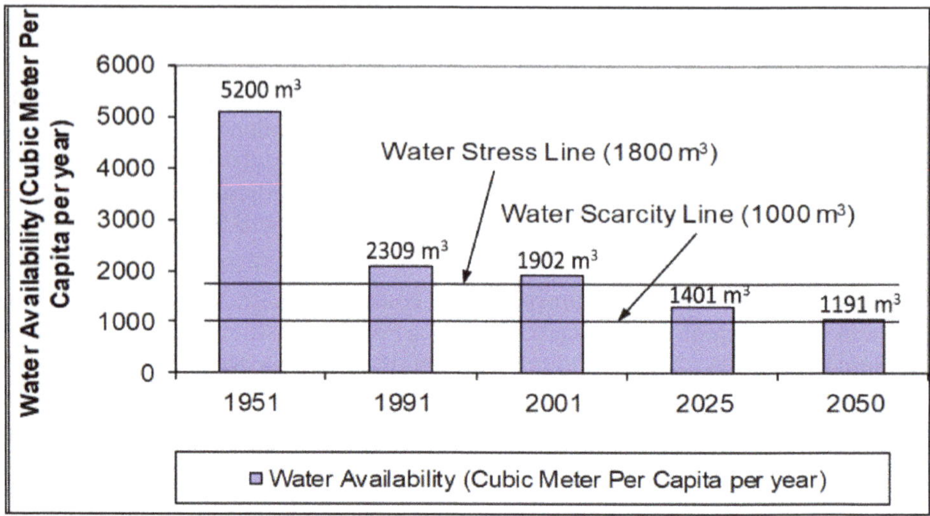

Figure 13.1: Per Capita Availability of Water in India.

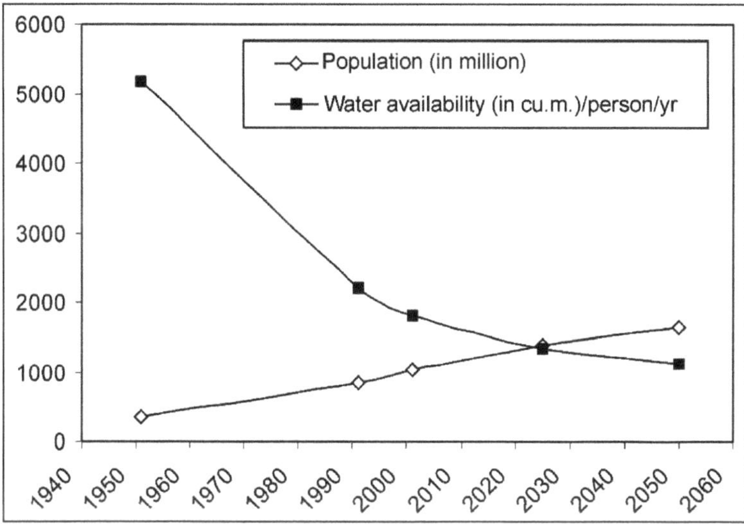

Figure 13.2: Graph Showing Population Expansion and the Declined Water Availability in India (After Mall *et al.*, 2006).

Added to the quantitative water scarcity, the deteriorating water quality is yet another major area of concern in India. The spatial incongruity has led to water quality deterioration due to poor management of natural resources and mismanagement of waste in cities and its fringes. On the other hand, the water quality of rural India is also deteriorating at an alarming level due to unscientific agricultural practices. According to a study carried out by World Resources Institute (Shiao *et al.*, 2015), 59 among the selected 632 heavily populated districts of India fulfill the drinking water standards set by Bureau of Indian Standard (BIS). The chemical parameters which breach the national standards include chlorine, fluoride, iron, arsenic, nitrate, and/or electrical conductivity. Further, the report says that about 130,600,000 people in India live in districts where at least one pollutant exceeded national safety standards in 2011 and more than 20 million people lived in the eight districts where at least three pollutants exceeded BIS permissible limits. Bagalkot in Karnataka is the most polluted, with five of six groundwater quality indicators at unsafe levels except for arsenic.

A study by Mekonnen and Hoekstra (2016) state that about two-thirds of the global population (4.0 billion people) live under conditions of severe water scarcity at least 1 month of the year. Nearly half of those people live in India and China. There is a need of mandating a clear cut policy on inter-State water sharing, since there have been several conflicts between states of Punjab and Haryana; Karnataka-Tamilnadu and Puducherry; Karnataka and Goa, Karnataka-Andhra Pradesh *etc.* These conflicts are escalating more often, causing wide-spread conflicts resulting in loss of lives and property, especially during low rain falls, shortage of water supplies and consequent droughts. Efficient management of available water even to the last drop, and innovating new technologies to harness unconventional sources of water are the only alternative in front of us.

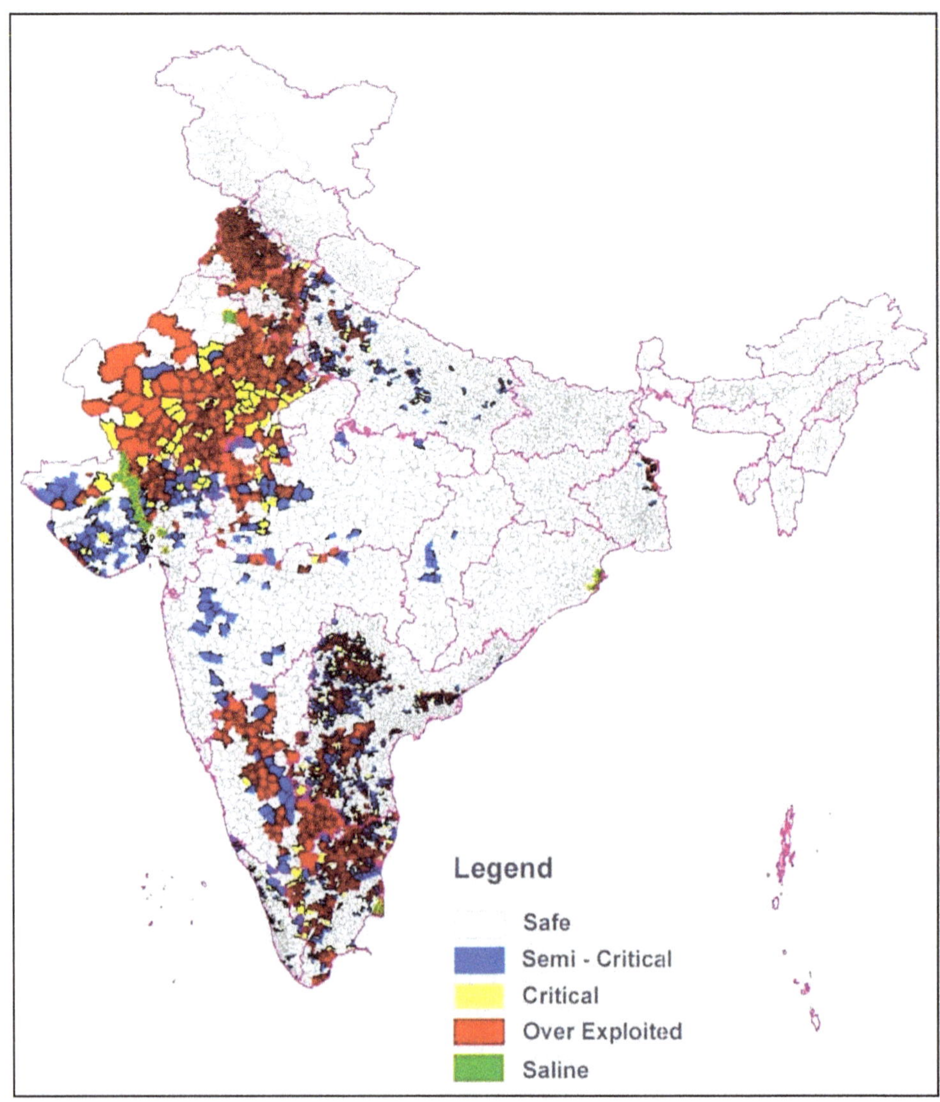

Figure 13.3: Quantitative Assessment of Groundwater Availability in India (after CGWB, 2016).

3. Strategies to Combat Water Scarcity

3.1 Innovative Approaches

3.1.1 Interlinking of Rivers

Interlinking of rivers is a dream project for Government of India and is quite challenging. The interlinking of rivers has two components: the Himalayan component and a Peninsular one. All interlinking schemes are aimed at transferring of water from one river system to another or by lifting across natural basins. The project will build 30 links and some 3000 storages to connect 37 Himalayan and Peninsular rivers to form a gigantic South Asian water grid. The canals, planned to be 50 to 100 meters wide and more than 6 meters deep, which would also facilitate navigation. The estimates of key project cost around 11.2 lakh crores (as estimated in 2014), to handle 178 km of inter-basin water transfer/per year, build 12,500 km of canals, create 35 GW of hydropower capacity, add 35 million hectares to India's irrigated areas, and create an unknown volume of navigation and fishery benefits. Similarly, 3700 MW would be required to lift water across major watershed ridges by up to 116 meters. With many positive effects, there are many socio-economic impacts. It is estimated that the network of canals extending to about 10,500 km would displace about 5.5 million people, who are mostly tribals and farmers (Vombatkere, 2003). In addition to this, submergence of forests and cultivable areas, displacement and resettlement and serious implications in terms of bio-diversity loss are the other challenges to face (Mehta and Mehta, 2013). Scientists also expressed concern that river diversion would bring drastic changes in the physical and chemical compositions of the sediment load, river morphology and the shape of the delta formed at the river mouth. However, in front of the larger benefit of water availability to the nook and corner of the country, the side effects may be less and can be handled with a detailed strategy. Socio-environmental, hydrological, geological and meteorological analysis is imperative in the large benefit of the country.

3.1.2 Diversion of West Flowing Rivers in Karnataka

The long range of Western Ghats in India is the place where most of the rivers of Peninsular India originate. This great escarpment forms a drainage divide where both east flowing as well as west flowing rivers originate. The intensity of rainfall in the Western Ghats in the 4 rainy months ranges from 5000 to 9000 mm with huge discharge of water into the Arabian Sea. The total water availability of water in the Western Ghats is 12254 billion cubic feet (Vaidhyanadhan and Subbarao, 2014) of which 6795 billion cubic feet of water flows towards west and the rest 5246 billion cubic feet to the east. Excluding the 40 per cent of the west flowing rivers as environmental flow which amounts to 2711 billion cubic feet, the excess 4084 billion cubic feet of water, if diverted towards the east, can solve the water problems of drought-prone Karnataka to a great extent. The excess water can also be shared with the neighbouring states. Sawkar *et al.* (2015), working on the hydrological feasibility of diverting the water from Sharavathi river, states that around 2000 billion cubic feet of water is drained to Arabian sea, which if utilized, can meet the requirement of a number of water-starved districts in the Karnataka state. Putty *et al.* (2015), working on the diversion of Aghanashini river to meet the water needs of Kolar, a drought prone district in Karnataka, projects the usage of 60 billion cubic feet of fresh water. However, the obstacle to implementing the diversion is the colossal budget requirement and the environmental consequences.

3.1.3 Cloud Seeding and Cloud Ionization

Cloud seeding and cloud ionization is a process of intensifying the precipitation through physical or chemical interaction with the clouds in the atmosphere. Cloud seeding is the chemical interaction which was invented by Vincent Schaefer in 1946 in New York. The chemicals generally used include silver iodide, potassium iodide and dry ice (solid carbon dioxide), calcium carbonate *etc.* Thesechemicals when delivered into the cloud through aircrafts, balloons *etc.* mimic the ice crystal and acts as condensation nuclei that can attract moisture in the cloud and form raindrops. This increased condensation and freezing releases a large amount of heat that makes cloud more buoyant and extends them sidewards and upwards. As clouds grow taller, their updraft increases and they draw in more moist air from the near surface and their size increases further. Enlarged clouds then encroach over several smaller clouds nearby and grow further, and hence the duration and quantity of rainfall will increase. Several cloud seeding efforts were taken up in Karnataka during 2017, and as claimed by the Government (of Karnataka) huge rainfall occurred in several drought prone regions of the State. However, systematic analysis of the data like the period of cloud seeding, the coordinates, the spatial extent covered, the amount of rainfall recurred, the density and humidity of clouds *etc.*, need to be carried out for an authentic documentation and to take this kind of initiatives further to help the drought prone areas.

Cloud ionization is the physical approach for enhancing precipitation in which collision covalence of the tiny cloud droplets are enhanced. The negative ions that are generated by a ground-based high-voltage corona-discharged wire array, move upwards by attaching themselves to the passing particles or aerosols in the air. Once they reach the level at which the clouds are forming, they become cloud droplets similar to the natural ones. But the artificially developed cloud droplets will have a negative electrical charge. Thus, when the positively charged natural cloud droplets collide with the negatively charged artificial cloud droplets, it increases the growth rate of rain droplets and ultimately results in heavy rain pour. The pilot studies of this technique (ART-ATLANT™, 2017) were initiated by Oman in association with Australian Rain Technologies in 2013 and they state that there has been an increase in the down pour of rain by 18 per cent (NIASRA 2017).

3.1.4 Tapping Deep Aquifers

The most critical aspect during droughts is ensuring at least safe drinking water for communities and livestock, especially in rural areas. The drought experienced from 2015 to 2017 in Karnataka showed that villagers struggled hard; even sometimes risking their lives to fetch drinking water. In such an emergency situation, the government can plan to exploit ground water from deep sources, normally at depths over 300 meters. Due to over exploitation and poor recharge (due to poor rainfall), most of the aquifers in Karnataka at shallow level (15–150 meters) are dried. Hence tapping water from deep aquifer could be a ray of hope for ensuring safe water during drought situations. Locating and developing deep aquifers require thorough geological and geophysical investigations and 'one well for one village' as a community well in the dire period of drought is a crucial step to be initiated by the concerned government.

3.1.5 Desalination

Desalinating sea-water can lend a great helping hand for India. Wit long coastline of 7,517 Kilometers, three fourth of India bordering with sea and with abundant sunshine in most part of the year are favourable conditions for tapping solar power and using it for desalination of sea water. Proven technologies, availability of resources and geographic conditions should help India to draft elaborate plans to ensure at least safe drinking water in the extreme events like droughts. Though laying pipes to transport desalinated sea water to drought prone areas is a challenging task, it will serve as a permanent solution and be greatly rewarded when the thirst of people is quenched with good drinking water.

Currently, there are many striking examples of drought-prone countries like Israel, Saudi Arabia, and Australia which have successfully implemented desalination programs. The recurring droughts have not only pushed Israel to desalinate sea water but also to adopt innovative technologies. Today 500 million cubic meters per year amounting to 15 per cent of Israel's water supply comes from desalinated sea-water. It provides about 35 per cent of Israel's drinking water which is expected to be about 70 per cent by 2050. From their sea-water desalination plants along the Mediterranean coast, elaborate pipe lines making use of both gravity and pumping have been laid for hundreds of kilometers with advanced quality control and command systems. At present Israel operates 5 large and 30 small desalination plants and this number is going to increase as per the plans by Ministry for Energy and Water Resources. Though Israel has the distinction of the country with largest per cent of desalinated water used in the world, several other countries facing water shortage are also using desalinated sea water. As per the International Desalinization Association (IDA Desalination Report 2016-17), 18426 desalination plants operate worldwide producing 86.8 million cubic meters of palatable water per day, providing access to 300 million people. The single largest desalination project is Ras-al-Khair in Saudi Arabia which produces about 1 million cubic meters per day. Their (Saudi Arabia) Al Khafiji desalination plant will be the world's first large scale solar powered desalination plant ready for commissioning which will supply for the entire city of Al-Khafiji. They have adopted novel technology of solar saline water reverse osmosis. The desalination plant in India near Chennai established in 2013 also produces about 36 million m^3 of water per year to partially meet Chennai's drinking water demand.

The cost of energy consumption by conventional desalination process costs about 3K Wh/m^3 which is not too high and is roughly equal to fresh water transported from large distances. Even this cost can be brought down by adopting solar saline water reverse osmosis process to the tune of 1 K Wh/m^3, as has been successfully demonstrated.

Today about 1 per cent of world's population is dependent on desalination water to meet their needs. But by 2025, as per the UN estimates, 14 per cent of world population will be facing acute water scarcity. And with unpredictable climatic variations and recurring droughts, India has to draw elaborate plans and implementation strategies to meet the drinking water shortage to the ever-swelling population. A desalination program can serve a great deal in this direction.

3.2 Intensifying Conventional Methods

3.2.1 Artificial Recharge

Artificial recharge is being implemented in India since pre-historic times to conserve rain water. There are numerous structures intact even today. The notable ones include Mahamandir Jhalara in Hodhpur, Pokhariyan ponds at Tikamgarh in Bundelkhand, Agrasen Ki Baoli in Delhi *etc.* The artificial recharge of water helps in conservation of water, diluting the contaminated and polluted water, and in reducing the risk of floods to some extent. However in the recent times, according to World Bank report (UNFCCC, 2007), the per capita storage of water is very meager as compared to the other nations (**Figure 13.4**).

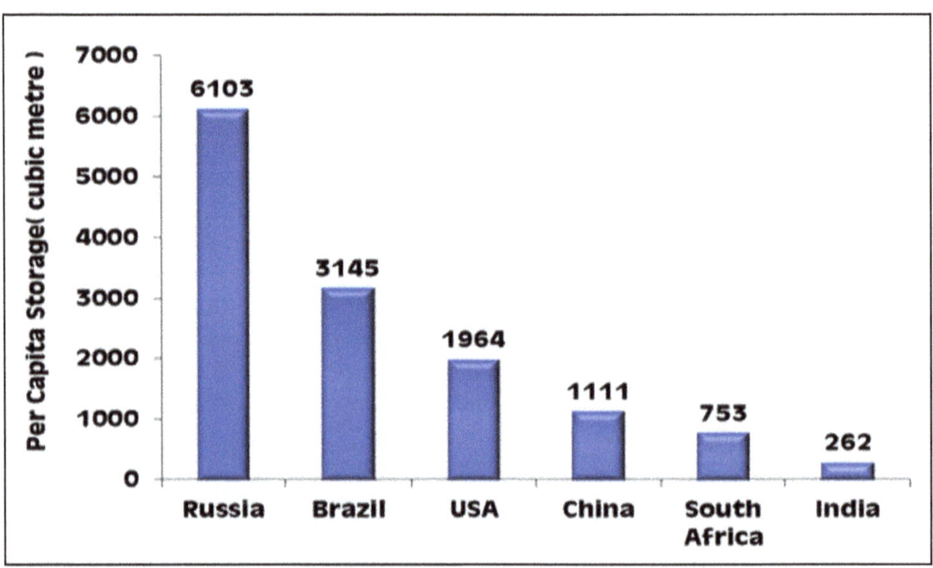

Figure 13.4: Per Capita Storage (Cubic meters) in different Countries.

The implementation of artificial recharge is site specific and hence the Central Ground Water Board (CGWB) has brought out guidelines for selecting the type of artificial recharge technique that suits the specific terrain conditions of the region. India has the capacity of harnessing 210,815 million cubic meters of water in various locations with a total areal extent of 795,850 sq. km in different states. In Karnataka, CGWB has identified around 747,007 Sq. Kms of recharge area wherein 11,367 million cubic meters of water can be harnessed. The site specific artificial recharge structures like 271 sub-surface dykes; 10,166 percolation tanks; 59,853 check dams *etc.* have been identified and being implemented (CGWB, 2016).

In recent times excess rainfall in urban areas has led to another problem called 'urban flooding'. Artificial recharge through roof top rainwater harvesting may bring a solution to this. For instance, Bengaluru is facing this problem in recent years. Erratic rainfall, poor drainage network, catchment of streams, choking of flow channels due to poor solid waste management and rapid urbanization made the city vulnerable to flooding. The city of Bengaluru receives an annual rainfall of around 965 mm with about 60 rainy days (one rainy day means a rainfall more than 2.5 mm). Conversely, Bangalore draws water from river Cauvery, about 100 Km away incurring huge financial costs. Water flows against gravitational force from Cauvery, and is pumped at various stages before reaching Bangalore. Thus, Cauvery travels a distance of 100 km to a height of 500 mt against gravity using 71 MW of electricity. Yet it faces an acute deficit of water supply against the total demand (Umamani and Manasi, 2013). In addition, residents also drill bore wells in order to ensure reliable supply of water. This has resulted not only in the rapid increase of the number of bore wells throughout Bangalore, but also overexploitation of ground water. Thus, rooftop rainwater harvesting is one of the potential solutions to combat water scarcity and also reduce the threat of the emerging disaster – urban flooding. Karnataka, with a potential of harnessing 167.33 Million cubic meters (MCM) of water through roof top rainwater harvesting may combat water scarcity effectively (WWAP, 2015).

3.2.2 Wastewater Treatment

The per capita wastewater generation by the Class-I and Class-II cities, representing 72 per cent of urban population in India, has been estimated to be around 98 liters/capita/day (lpcd) (CPCB, 1999). As per CPCB estimates, the total wastewater generation from Class I cities (498) and Class-II (410) cities in the country is around 35,558 and 2,696 million liters/day(MLD), respectively. The installed sewage treatment capacity is just 11,553 and 233MLD, respectively, which still leaves a gap of 26,468 MLD in sewage treatment capacity. Maharashtra, Delhi, Uttar Pradesh, West Bengal and Gujarat are the major contributors of wastewater (63 per cent; CPCB, 2007). Further, as per the UNESCO and WWAP estimates (Van-Rooijen *et al.*, 2008), the industrial water use productivity of India (IWP, in billion constant 1995 US$ per m^3) is the lowest (*i.e.* just 3.42) and about 1/30th of that for Japan and Republic of Korea. It is projected that by 2050, about 48.2 Billion Cubic Meters (BCM) about 132 billion litres per day of wastewaters (with a potential to meet 4.5 per cent of the total irrigation water demand) would be generated, thereby further widening this gap (Bhardwaj, 2005). Thus, overall analysis of water resources indicates that in coming years, there will be a twin-edged problem to deal with reduced fresh water availability and increased wastewater generation due to increased population and industrialization. Conversely in Singapore, wastewater becomes new water *i.e.* potable water which they call NEWater. Wastewater treatment supplies around one third of the country's water demand, and that number is expected to grow to more than half by the year 2060. There are now four purification plants across Singapore producing 430 million liters of new water a day. The majority of what's produced is consumed by industry or by big cooling facilities. The rest is combined with nutrient-rich reservoir water, purified again and filled into bottles. Hence, there is a need to take this type of initiatives in India which helps in sustainable development and management of water resources.

3.2.3 Water Vapour Harvesting using Nanotechnology

Nowadays nanotechnology is widely used in water treatments. The nanomaterials such as antimicrobial nanomaterials, carbon nanotubes (CNTs), nanosorbents, dendrimers, self-assembled monolayers on mesoporous silica (SAMMS), Single Enzyme Nanoparticles (SENs) are being used for the purification of water (Dhakras, 2011). Nanotechnology is also used for harvesting water vapour from air. The researchers from the School of Chemistry and Australian Institute for Nanoscale Science and Technology (AINST) being inspired by a species of desert beetle that uses nanoparticles to tap water from the atmosphere, which can be used in the circumstances of drought, emergency or isolation from the main water grid.

4. Conclusions

Ensuring adequate fresh water to communities and livestock has been emerging as a biggest challenge to India in the face of climate change and the consequent recurring droughts. History has clearly shown that millions of people in India have perished due to famines, the acute form of droughts. Thanks to the advent of science, that we are in a better position to combat these natural hazards. Yet, the great advancements in handling these situations are jeopardised by the shrinking water resources and swelling population. Hence, measures have to be initiated on war-footing with an integrated approach of educating masses on the importance of water; it's conservation; judicious use; harnessing various sources; employing scientific know-how to trap the various sources of water including recycling wastewater; rationalizing policies to prioritise the water resources issues, especially in reducing water consumption for irrigation; and adopting the UN Policy that 'access to fresh and safe water to all' as an universal human right (UNDP, 2006). Underprivileged communities in rural India should be brought to awareness that any natural hazard like drought should not be viewed as a curse on them, but to demand from the government to supply safe water. In a nutshell, a large-scale awakening is required at all levels to combat the hazards of water scarcity.

Acknowledgements

The authors would like to thank Dr. B.N. Rashmi, Project Scientist, Karnataka State Natural Disaster Monioring Center, Bangalore for providing useful information about the drought situation in Karnataka.

References

1. ADB-PIK, 2018. A report: Asian Development Bank: A region at risk. The human dimensions of climate change in Asia and the Pacific.

   Accessed on 15 Jan 2018 https://www.adb.org/sites/default/files/publication/325251/region-risk-climate-change.pdf

2. Amarasinghe, U.A., Shah, T., Turral, H., and Anand B.K., 2007. India's water future to 2025-2050: Business as usual scenario and deviations. Report no. 123, International water management Institute. p.42, ISSN: 1026-0862.

3. ART-ATLANT™. Australian Rain Technologies, Sydney, NSW, Australia.

   Website report https://www.australianrain.com.au/

4. Bhardwaj RM. 2005. Status of wastewater generation and treatment in India, IWG-Env Joint Work Session on Water Statistics, Vienna, 20-22 June 2005.

5. CGWB 2017. Central Ground Water Board Year Book 2016-17. Report of the groundwater availability of India, Faridabad. p.90.

   http://www.indiaenvironmentportal.org.in

6. CGWB 2016. Central Ground Water Board Year Book 2015-16. Report of the groundwater availability of Karnataka. p. 194.

   http://cgwb.gov.in/gw_profiles/st_Karnataka.htm

7. CPCB 1999. Status of water supply and wastewater collection treatment and disposal in Class-I Cities-1999, Control of Urban Pollution Series: CUPS/44/1999-2000. Central Pollution Control Board, India.

8. CPCB 2007. Evaluation of operation and maintenance of sewage treatment plants in India-2007, Control of Urban Pollution Series: CUPS/68/2007. Central Pollution Control Board, India.

9. Dhakras, P. A. 2011. Nanotechnology applications in water purification and wastewater treatment: A Review. International Conference on Nanoscience, Engineering and Technology, Tamilnadu, p.285-291.

   https://doi.org/10.1109/ICONSET.2011.6167965

10. Eakins, B.W. and Sharman, G.F,.Volumes of the World's Oceans from ETOPO1, NOAA National Geophysical Data Center, Boulder, CO, 2010.

11. Gupta, A.K., Tyagi,P., and Sehgal, V.K. 2011. Drought disaster challenges and mitigation in India: strategic appraisal. *Current Science,* 100: 1795 – 1806.

1. IDA Desalination Yearbook, 2017. Annual IDA water desalinization report 2016-17. Topsfield, MA 01983, USA.

   https://www.globalwaterintel.com/static_file_containers/112

13. Mall, R.K., Gupta, A., Singh, R., Singh, R.S., and Rathore, L.S. 2006. Water resources and climate change: An Indian perspective. *Current Science,* 90:1610 – 1626.

14. Mehta, D., and Mehta, N.K., 2013. Interlinking of Rivers in India: Issues and Challenges.Geo-Eco-Marina. p. 138-143.

    https://www.geoecomar.ro

15. Mekonnen, M. M., and Hoekstra, A. Y. 2016. Four billion people facing severe water scarcity. *Science Advances,* 2: e1500323.

    http://doi.org/10.1126/sciadv.1500323

16. NIASRA 2017. 2016-Oman rainfall enhancement. University of Wollongong. Pp.60.

    weblink: www.australian.rain.com.au

17. Obasi, G.O.P., 1994. WMO's role in the international decade for natural disaster reduction. *Bull Amer Meteor Soc.,* 75: 1655-1661.

18. Oldenborgh, G.J., Philip, S., Kew.S., Weele, M.V., Uhe, P., Otto, F., Singh, R., Pai, I., Cullen, H., and Rao, K.A., 2018. Extreme heat in India and anthropogenic climate change. *Journal of Natural Hazards Earth System Science,* 18: 365-381.

19. Putty, Y.R., Bhat,V., Mohan, V.S., and Sushma, R., 2015. A Study on the hydrological feasibility of diversion of the west flowing Aghanashini. p. 6.

http://www.kscst.iisc.ernet.in

20. Sawkar, R.H., Rudraiah, R. and Madhav, 2016. Transfer of surplus water from west to east flowing rivers of Karnataka: Management of surface water and groundwater in drought prone areas. *Geol. Soc. India Spec. Publ.*, 5: 76-85.

21. Shiao, T., T. Luo, D. Maggo, E. Loizeaux, C. Carson, and Shilpa Nischal. 2015. India Water Tool. Technical Note. Washington, D.C.: World Resources Institute.

Available online at: www.wri.org/publication/india-water-tool.

22. Umamani, K.S., and Manasi, S, 2013. Rainwater harvesting initiative in Bangalore City: Problems and Prospects. Working paper–302, Institute for Social and Economic Change, Bangalore. p.24 ISBN 978-81-7791-158-9.

23. UNDP 2006. United Nations Development Programme Annual Report, 2006–Global partnership for development. p. 40.

http://www.undp.org

24. UNEP 2016. A snapshot of the world's water quality: towards a global assessment. United Nations Environment Programme (UNEP), Nairobi.

25. UNFCCC 2007. United Nations framework convention on cimate change. Report of the Conference of the Parties on its thirteenth session, held in Bali from 3-15th December 2007. p.57

26. Vaidhyanadhan, R., and Subbarao, K.V., 2014. Landforms of India from topomaps and images. Geological Society of India, Bangalore. ISBN 978-93-80998-02-2.

27. Van-Rooijen D.J., Turral H., Biggs T.W., 2008. Urban and industrial water use in the Krishna Basin', irrigation and drainage. DOI: 10.1002/ird.439.

28. Vombatkere, S.G., 2003. Interlinking: Salvation or folly?

Weblink: http://www.indiatogether.org

29. World Bank 2010. World development indicators database. released on 1st July 2011

https://siteresources.worldbank.org/DATASTATISTICS/Resources/POP.pdf

30. WWAP 2015. The United Nations world water development report: Water for a sustainable world: facts and figures. United Nations World Water Assessment Programme (WWAP). United Nations Educational, Scientific and Cultural Organization, Paris, France.

31. WWAP 2017. The United Nations World Water Development Report: Wastewater, the untapped resource. United Nations World Water Assessment Programme (WWAP). United Nations Educational, Scientific and Cultural Organization, Paris, France.

Chapter 14

Rural Communities Response to Climate Change in Malawi: The Case of Mzimba District

G.R. Phiri

University of Ekwendeni, P. O. Ekwendeni, Malawi
Centre for Open and Distance Learning (CODL), Mzuzu University, Private Bag 201, Luwinga, Mzuzu 2
Malawi Union of Academic and Non – Fiction Authors (MUANA), P.O. Box 31942, Blantyre, Malawi
Centre for Development and Information in Southern Africa (CEDRISA), Lilongwe, Malawi
E-mail: enkupeniconglomerate@gmail.com, gr_phiri@yahoo.co.uk

ABSTRACT

Climate change poses multiple challenges to the Government of Malawi just like Governments of her sister developing countries. Experience has shown, however, that high – tech strategies lined up by government can trickle down to beneficiary rural communities only where and when they have been merged with rural communities knowledge and practices. The present paper deals with hi-end technologies for alleviating impacts of climate change in Mzimba District, which are largely inappropriate given the missing link between communities of practice i.e., one represented by rural communities' indigenous knowledge about their environment and the biodiversity therein and that operating in the scientific arena. To this effect, the two communities of practice need to be brought under the same roof. The merger has many benefits, namely, it provides a platform for comparing notes and sharing experience using real time; rural communities quickly amass scientific knowledge; rural communities' acquaintance with scientists will avail to them an opportunity for their hidden literary treasures to be documented and preservation of riverine vegetation and reforestation of river banks in the South Rukuru riparian area is anticipated to yield multiple benefits including the rival of fisheries locally known as Kazuni. In a significant way, a revival of riverine vegetation in the riparian area will increase the intake of carbon dioxide.

Keywords: *Accentuated relief, Climate change, Cosmetic laws and policies, Erratic rainfall, Planting rains, Ramifications, Riparian area, Reforestation, Resilience and Response.*

1. Introduction

Climate change poses multiple challenges to the Government of Malawi just like governments of Malawi's sister developing countries. There is a lot of literature that proves that the Government and her development partners earnestly seek to avert the impacts of climate change. It is the conviction of this author, however, that high – tech strategies lined up by government can trickle down to beneficiary rural communities.

1.1 Location, Physiography and Drainage

Malawi stretches from latitudes 9° to 17°S and longitudes 33° to 34°E. The country has accentuated relief which is most pronounced in the Northern Region where Mzimba District lies. Mzimba District lies on a 1,100-metre-high gentle plain (Mzimba Plain). The Mzimba Plain is bordered by a 1,500 to 1,860-metre-high Viphya Plateau system on the eastern edge and a 2,000 – metre – high Nyika Plateau system on the northern edge. The Plain extends westwards into Zambia and beyond. Meanwhile, the Viphya and Nyika Plateau systems are part of the southern Tanzania mountain block, save that the two blocks in Malawi are separated from that in southern Tanzania by the Lake Malawi rift. The lake Malawi rift floor lies at a low 550 metres.

Furthermore, the section of Mzimba Plain, which Mzimba District occupies, is drained by South Rukuru River – the longest river in the Northern Region of Malawi. The river runs from the southern end of the Viphya Plateau system to the edge of the Nyika Plateau system on its northern tip. The river makes a long loop thus flowing westwards initially, then northwards and eventually eastwards. It separates the Viphya Plateau system from the Nyika Plateau system. It empties into Lake Malawi. Mzimba District, therefore, coincides with the South Rukuru riparian area.

1.2 Climate

Malawi has accentuated relief of this; the country experiences a modified Sudan climate, which is characterized by seasonal rainfall. The rains are produced by the migratory inter-tropical convergence zone (ITCZ). The ITCZ draws in and converges the Northeast Trade Winds, Southeast Trade Winds and the Congo Airstream. The Mzimba Plain lies on the leeward side of the two trade wind systems. On the contrary, the Plain lies on the windward side of Congo Airstream. Naturally, the amount and distribution of rainfall across the Mzimba Plain heavily relies on which of the three air masses is dominant and for how long.

2. Expressions of Climate Change in Malawi

Climate change in Malawi has reached such a high magnitude that it creates a large number of "Climate-refugees" (*i.e. persons internally displaced by severe effects of climate change*) year after year. In a nutshell, rural communities in Malawi operate in very stressful circumstances characterized by pervasive poverty that is aggravated by a single, seasonal export crop economic base, a low electricity consumption rate, high levels of illiteracy, a youthful population and a governance system that is characterized by cosmetic laws and policies. The term cosmetic laws and policies is used here to imply governments overt failure to support rural communities to build their resilience to climate change regardless of an adequately articulated legal and policy framework. Given this background, rural communities feel the full brunt of climate change – induced severe weather conditions, such as-

- ✰ Erratic rainfall, which pours in brief periods of very high intensity rains. Incidences of high intensity rains lead to extensive flooding. On other hand, the brief, high intensity rains are punctuated by long dry spells. Their effect is obviously the direct opposite of floods.
- ✰ Progressively low annual rainfall, which culminates into correspondingly low water tables, dwindling wetlands, dry river courses, and low lake levels.
- ✰ Substantially high food insecurity.
- ✰ "Dry" taps and extremely low electricity generation rate.
- ✰ High incidences of heat "islands", which translate into localize, intense "dry or wet" whirlwinds.

3. Ramifications of Climate Change for the Resilience of Rural Communities

While realizing that there is a thin line between the manner in which climate change expresses itself and its effects, it is necessary to rebrand the effects as ramifications. Climate change in Malawi has a very wide diversity of ramifications. Although this is common knowledge, the major challenge for the country concerns choosing among an equally wide range of plausible avenues for handling the ramifications to alleviate their toll on humans and the environment.

Basically:

- ✰ All ramifications of climate change need urgent attention. Furthermore, to be of any consequence, they need to be tackled simultaneously.
- ✰ On one hand, rural communities have a rich, but undocumented treasure of indigenous knowledge about the ecology, pedology and hydrology of their environment. On the other hand, the country is awash with documented, scientific knowledge in the same fields.
- ✰ There is a missing link between communities of practice that should otherwise be complementary. The two communities of practice are:

a. One that is based on the indigenous knowledge of rural communities, and

b. another that is based on scientific knowledge.

4. An Insight into Remedial Courses of Action

It is obviously difficult to find remedies to diverse ramifications of climate change especially given the fact that they are multi–faceted as well. Noted that current paper suggests a strategy that has a cumulative or snow ball effect, and to remain focused, a three tier response to climate change, namely, weather forecasting, land use planning and water resources management is proposed. In nutshell, traditional practices in respect of weather forecasting, land use planning and water resources management are based on painstaking, aural tradition while those based on science depend heavily on experiments and records.

4.1 Weather Forecasting

a. Use of Plants

Plants are used to estimate the onset of planting rains, planting rains comprise of 25 mm plus rainfall incident at the beginning of the rainy (planting) season. The key plants are:

- ☆ Bulbs *e.g.* ***nkhorwa*** in vernacular (*Hippeastrum* sp.): The wild onion sprouts, its stem 14 days prior to the onset of rains. The photos below illustrate the plant. see **Figure 14.1.**

Figure 14.1: *Nkhorwa* (*Hippeastrum* sp.) (a) *In the Dry Season* and (b) In the Rainy Season.

- ☆ Blood lilies (*e.g. Haemanthus species*): The blood lilies are an elongated tuber whose stem sprouts 28 days prior to the onset of rains. Photos below illustrate the tuber, see **Figure 14.2**.
- ☆ Corns (*Colocasia* sp.): One of the corns which are used to estimate the onset of rains is called *Dema*. Dema is basically a *Colocasia species*. It is a large tuber whose multiple stems sprout 42 days prior to the onset of

Figure 14.2: Blood Lilies (*Haemanthus* sp.) (a) During the Dry Season and (b) During the Wet Season.

rains. Using simple proportion the three plants point to the same day as the onset of rains.

b. Use of Insects

An additional avenue for estimating the onset of rains is making an observation of insects. The most reliable insects are ***a cicada*** and **grasshoppers**.

- ☆ A Cicada (*Megapomponia imperatoria*): A cicada is a large insect which makes a continuous loud noise. It has a short life span. It emerges in forests from around 42 days prior to the onset of rains. Immediately the rainy season gets stabilized, the cicada phases out (**Figure 14.3**).

Figure 14.3: A Cicada, *Megapomponia imperatoria*

- ☆ *Jaghayagha* (vernacular): This is a large grasshopper which also makes a sharp continuous noise by rubbing its hind legs against its body. It produces the continuous noise intermittently until the rainy season stabilizes.

- *Ntheputepu* (vernacular): ***Ntheputepu*** is a giant but slender green grasshopper which emerges around 42 days prior to the onset of rains.

c. Use of Birds

Birds are used to estimate the onset, amount and distribution of rainfall. The most reliable birds in this category are migratory birds and one known as ***Jongwe*** *(a jail bird)*.

- Migratory birds around lake Chilwa (Lake Chilwa basin is the hubs of climate adaptation programme in the country. Its major objective is to raise the ecological and social resilience of the environment and communities, respectively, in Lake Chilwa wetland), one of two Ramsar sites in Malawi, fly out of country around 30 days prior to the onset of rains.
- ***Jongwe*** *(a jail bird)* is a medium – sized bird with a long beak. It lives mostly on nectar. The bird builds a cotton – lined, crescent – shaped nest 35 days prior to the onset of rains. It positions its nest in such a way that the entrance faces away from the direction of the dominant air mass among the three airstreams listed in the section on climate. The nest's position is aimed at protecting the eggs and the chicks from rain drops **Figure 14.4.**

Figure 14.4: Nest of Jongwe, Jail Bird.

d. Scientific Bases for Rural Communities Knowledge and Tips of Food Security

Much as they are unaware, rural communities indigenous knowledge is [illegible]ed on scientific principles. The first scientific principle is sensitivity to a progressive pick up in atmospheric relative humidity. If one starts with plants, one notes that they are activated by a threshold level of atmospheric humidity. Given the seasonal nature of rainfall in the country, the tubers listed in the paper sprout the stems prior to the rains in a bid to produce seeds using the food they manufactured and preserved in the previous rainy season. The leaves that follow manufacture plant food for preservation and use at the start of the subsequent rainy season. On their part, insects emerge on Mzimba plain on the verge of the rainy season driven by strict levels of atmospheric humidity. Here, the eggs which were laid in the prior season hatch when an appropriate relative humidity level has been reached. Just like plants, the insects are so perfectly adapted that their eggs hatch at a time in the year when there is sufficient food (grass, leaves and nectar) for the offspring.

Meanwhile, ***Jongwe*** [*(vernacular) jail bird*], whose population is big on Mzimba plain, reveals additional information to the onset of rains. By positioning its nest strategically, it enables rural communities to estimate the amount and distribution of rains prior to their onset. In a year when the nest faces westwards, low and erratic rainfall is anticipated. The appropriate courses of action in such a year include the following: planting crops with low moisture requirements, drought tolerant crops such as millets and sorghum, pumpkins, watermelons and legumes, namely, running beans, ground beans and soybeans. Root crops of all manners are also recommended for planting in a low, unevenly distributed rainfall.

4.2 Land Use Planning

Science – based land management is alternatively known as *land evaluation* or *land use planning*. Soil surveys and land capability classification form the backbone of land management. Additionally, scientific principle pertaining to the choice of appropriate soils for root crops is also essential. Rural communities need to accurately locate the natural "habitats" of wild root plants listed in an earlier section of this paper. *In fact, in the very distant past, scouts used wild root plants to choose sites for their ever – shifting villages.*

In a year when the nest faces east or north, adequate, evenly distributed rains are anticipated by rural communities. During such a season, all crops, regardless of their moisture requirements, are recommended to be planted. Meanwhile, crops with low moisture requirements are planted towards the end of the season.

Regardless of the amount and distribution of rainfall, age-old food preservation technologies should be revived. They include sun drying and packaging of leaves of wild plants such as Beans, pumpkins, and mushrooms. Leaves of beans (*i.e.* ordinary and running beans) and mushrooms should be collected from forested patches and peelings of sugar – rich fruits including ***mbula or maula*** (vernacular).

4.2.1 An Additional Application of Scientific Principles

Much of the human body is sensitive to variations in atmospheric humidity, even though humans cannot be comparable to plants, insects and birds as articulated in this paper. Since procuring and distributing hygrometers and Stevenson's Screens is out of question, a low cost, easy to use "**gadget**" is a 900 cm^2 porous ceiling board. The piece should be immersed in a solution of cobalt chloride and water until it is fully soaked. Once it is taken out, the piece should be left hanging under an adequately aerated shed. During the rest of the dry season, the cobalt – laden ceiling board will assume its normal colour. It will turn purple (the colour of wet cobalt) immediately relative humidity starts picking up. The communities should cross check with tips "provided" by plants, insects, grasshoppers and birds. With the passage of time, they will be able to make intelligent estimates of atmospheric humidity as the onset of rains draws near.

4.3 Water Resources Management

The final plausible response to climate change to be recommended pertains to tips on water resource management. In the distant past, scouts looking for good land to locate new villages were equipped with tips on how to identify aquifers. *Katope* (vernacular) was used to locate aquifers. The scientific explanation for this is that the tree anchors its tap root in the aquifer.

On the other hand, the African teak (*Pterocarpus angolensis i.e. mlombwa or mbira* in vernacular) sups up a lot of water during the rainy season, holds it up in its trunk only to release it back towards the ground surface during the dry season. Where it grows along river courses, *the African teak* recharges the said rivers during the dry season. One recommendation to rural communities on Mzimba Plain is to identify aquifers using *Katope trees*. Once this has been done, the communities should be encouraged to establish African teak (*Pterocarpus angolensis*) – rich woodlots over them.

Bearing these two initial benefits in mind, preserving trees and reforestation of river courses in South Rukuru riparian area can be readily accepted by the rural communities because they will be building on their age – old community of practice.

5. Conclusion

This paper address hi-level technologies for alleviating impacts of climate change in Mzimba District are largely inappropriate given the missing link between communities of practice *i.e.* one represented by rural communities' indigenous knowledge about their environment and the biodiversity therein and that operating in the scientific arena. To this effect, the two communities of practice need to be brought under the same roof. The merger has many benefits, namely,

a. It provides a platform for comparing notes and sharing experience
b. Rural communities quickly amass scientific principles because-
    - ☆ All they need to do is to apply scientific explanations to their very rich indigenous knowledge
    - ☆ The suggested strategies are mostly free of any kind of costs. Where costs are involved, they are negligible. What is overwhelmingly pleasing though, is that the strategies for reviving natural vegetation in the South Rukuru riparian area provides a rare avenue for enhancing capacity for community – based natural resources monitoring that informs practical decision – making using rural, age- old traditional

cum cultural structures in such a way that government and other formal governance structures are unable to.

- ✰ The suggested strategies fill the gap created by extensive deforestation in most parts of Mzimba District. Extensive deforestation has removed most of the plants, insects, grasshoppers and birds on which rural communities base their decisions.

c. Rural communities' acquaintance with scientists will avail to them an opportunity for their hidden literary treasures to be documented. The hidden treasures in question include heavily guarded family or community – specific coping skills some of which are based on myths, beliefs and customs.

d. Preservation of riverine vegetation and reforestation of river banks in the South Rukuru riparian area is anticipated to yield multiple benefits including the rival of fisheries locally known as *Kazuni*. In a significant way, a revival of riverine vegetation in the riparian area will increase the intake of carbon dioxide.

References

1. Berrie, G. K., Berrie A and Eze, J.M.O. 1987. Tropical Plant Science. John Wiley and Sons, Inc. pp. 42 – 46, New York, USA

2. Gondwe C. 2002. Community – based natural resources resource management initiatives take root in Malawi. The success story of Kam'mwamba community in the conservation of trees and forests. *Future forest technology, environment, economics and administration* 2: 5 – 6.

3. Phiri, G.R. 2012. Lifting the lid on HIV/AIDS and Tuberculosis in Malawi. In Taylor, Neil *et. al.*, Health Education in Context – An International Perspective on Health Education in Schools and Local Communities. Rotterdam Sense Publishers – pp. 90

4. Phiri, GR. 2017. Enhancing lightning hazard mitigation through traditional customs and religion in Malawi. In Hole, R.L. and Ataremwa, Edmund (Editors): Lightning Impacts in Developing Countries of Africa and Asia Centre for Science and Technology of the Non – Aligned and Other Developing Countries (NAM S&T Centre) – pp. 18

5. Tisanthule – A Zodiac Television programme which analyzes the discrepancy between Malawians' unparalleled creativity and their failure to USE its products to their advantage.

Chapter 15

Understanding Connections between Climate, Extreme Weather, Air Quality, and Health with a Glance at Sri Lanka: A Review

H.K.W.I. Jayawardena

Department of Physics, Open University of Sri Lanka, Nawala, Nugegoda 10250, Sri Lanka
E-mail: hkjay2@ou.ac.lk

ABSTRACT

Climatic changes result in an increased occurrence of extreme weather events. Air pollution is exacerbating the situation, provoking even more health issues. This Review investigates the health impacts and potential synergies between extreme weather events and air pollution, summarizing some prominent studies published between 2014 and 2017. Additionally, extreme weather temperatures and air pollution with particulate matter was examined using past observations. The literature study reveals possible synergetic effects of air quality and meteorology that would be useful in improving weather forecasts related to health impacts. Annual concentrations of particulate matter in the Colombo city area exceeded the permissible levels for both Sri Lanka and the World Health Organization. Extreme warm day percentages increased during the last few decades, indicating changed climate and hence weather in Sri Lanka. The study reflects the country's possible combined effects in heat-related weather extremes and poor air quality which can cause numerous health effects. The study should develop to uncover hidden interactions between pollutants and weather extremes considering health consequence. Such efforts will improve the extreme weather mitigation systems regulating the human health burden. Moreover, weather forecasts can be enhanced to give extra warnings to specific vulnerable populations.

Keywords: *Air pollution, Extreme weather, Climate change, Health impact, Sri Lanka, Particulate matter, Tx90p.*

1. Introduction

Human health is highly sensitive to climate variations. Extreme climate events have become more frequent as a result of climate change, posing increasing threats to public health. Moreover, air pollution, a by-product of increasing urbanization and industrialization, is worsening the situation, changing the climate and increasing the health burden. Therefore, investigating the combined threat can be a useful guidance to mitigate climate change related issues.

IPCC (*Intergovernmental Panel on Climate Change*) Assessment Reports (AR), AR4 (IPCC, 2007) and AR5 (IPCC, 2013) reported evidence for an increase in greenhouses gases (GHGs) and aerosols during past decades that affect the Earth's radiation balance and accelerate climate change with frequent extreme events (IPCC, 2013).

The WHO (*World Health Organization*) also indicates an increase in the urban air pollution level by 8 per cent during between 2008 to 2013. Their recent media reports indicate that more than 92 per cent of people living in the urban areas in low and middle income countries are exposed to poor air quality (WHO, 2016). According to the BreatheLife campaign, which is a program developed by WHO in partnership with the CCAC (*Climate and Clean Air Coalition*), the annual average concentration of particulate matter (PM) 2.5 particles that have a diameter of less than 2.5 mm in Colombo is 36 $\mu g/m^3$ while 27 $\mu g/m^3$ for whole country (BreatheLife, 2017). These amounts exceed many folds to the WHO annual mean safe level (10$\mu g/m^3$), indicating the increased health risk associated with poor air quality.

Fine PM2.5 exposure has been cited as the fifth largest mortality risk factor in 2015 and responsible for 4.2 million deaths from various cardiovascular and cerebrovascular diseases including cancers (SOGA, 2017). It is also revealed as a major cause of premature death in South Asia (DeSario *et al.*, 2013; Jain *et al.*, 2017; Ghudeet al. 2016). It is reported in health statistics that ischemic heart disease, cerebrovascular disease and chronic obstructive pulmonary diseases (COPD) has increased in Sri Lanka by 11.9, 4.2 and 11.6 per cent respectively during the 2005 to 2016 period (IHME, 2017). There could be several factors behind these figures other than air pollution, but it is worthwhile to explore the effect of air quality on such health issues.

According to the Central Environmental Authority (CEA) of Sri Lanka, the ozone (O_3) level also significantly exceeds the recommended level by the WHO (Jayathilake and Jayarathne 2016). This ground-level ozone can affect the respiratory system and can increase morbidity and mortality in some sensitive groups of population. Considering all these facts on climate, air pollution and health impacts, a priority should be given to incorporate interactions related to air pollution when improving extreme weather forecasts. To address this need, as an initial approach, this review explores the scientific evidence of interactions between extreme weather events and air quality, based on research work published from 2014 to present. Further, a basic study was carried out to identify the possible climate-air quality situation in Sri Lanka using past air quality observations and climatological data.

The final aim of the study is to develop a framework of tools including forecasting methodology to improve the preparedness for and response the adverse health impacts of air pollutants considering the synergies with extreme weather events.

2. Materials and Methods

The review followed studies of direct and indirect interactions of climate, extreme weather, air quality and health published during 2014 to 2017 period in journals and on the web. The study used annual concentrations of PM10 in the Colombo city area (Fort railway station) from between 1998 and 2004 to examine its annual variation. The data were obtained from Central Environmental Authority of Sri Lanka.

To investigate the occurrence of thermal threshold escalation as an evidence of extreme climate in Sri Lanka, the Tx90p Climate Index is used in the study. Tx represents daily maximum temperature while Tx90p is warm day frequency and it is the daily maximum temperatures above 90th per centile between 1961 and 1990. Annual extreme warm day percentages (TX90p) for 8 stations (*Colombo, Rathnapura, Batticaloa, Anuradhapura, Badulla, Hambanthota, Kurunegala, NuwaraEliya*) has been investigated in this study. This TX90p data were computed by the Expert Team on Climate Change Detection and Indices (ETCCDI), Canada using number of global climate models (Sillmann *et al.*, 2013a, b).

3. Results and Discussions

3.1 Information Analysis

The problem of air pollution is twofold. Firstly, increased pollutants trap the sunlight and cause global warming and climate change. Secondly, warmer climate will lead to more severe air pollution. Therefore, air pollution and climate change are closely connected in both directions.

Long-term aerosol pollutants such as soot, dust and other small particulate matter can affect cloud formation and rainfall, changing the cloud height and thickness and modifying the precipitation frequency and intensity, exacerbating extreme weather conditions (Li *et al.*, 2011). A study based on a multi-scale global climate model has revealed that the Asian air pollution had strengthened storm clouds by delaying release of precipitation and making the clouds live longer and grow bigger (Wang *et al.*, 2014). Nevertheless, weather and climate influences air pollutant dispersion and transportation and chemical transformations that affect air quality in various ways (Leelõssy *et al.*, 2014). Temperature, humidity and wind also have significant influences from increased air pollution (Wang *et al.*, 2014, Zeng and Zhang 2017).

Numerous studies have examined the impact of extreme weather conditions on health, based on air quality. A very recent study has discovered an increased risk of heart attack associated with small particulate pollution with certain blood types, which can be an epidemic effect in a situation of extreme weather condition (Intermountain Medical Center, 2017). **Table 15.1** summarizes some prominent literature published during last 5 years that considers the interactions between

Table 15.1: Studies Addressing the Research Question – Interactions between different Types of Weather or Climate Extremes and Air Pollution Leading to Adverse Health Effects

Reference Region and Period	*Extreme Weather*	*Study Linked to Air Quality*	*Health Effect*	*Indices/Factors/Methods Used*	*Results*
Zhang B. *et al.*, 2017 USA (2017)	Extreme O_3 and PM 2.5 events. Extreme weather elements - Maximum temperature, minimum relative humidity and wind speed.	A quantitative study of the severe air pollution events and extreme weather events based on historical observations	Not considered	Extreme O_3 and PM 2.5 events daily maximum temperature (T_{max}), minimum relative humidity (RH_{min}), and minimum wind speed (V_{min}), using the location-specific 95th or 5th per centile threshold derived from historical reanalysis data	Annual O_3 and PM 2.5 extreme days were highly correlated with T_{max} and RH_{min} while positively (negatively) correlated with Vmin in urban (rural and suburban) stations.
D'Amato *et al.*, 2016 Australia (2016)	Heavy thunderstorms	High pollen grain concentrations in the air during the thunderstorms, inducing asthmatic reactions in patients	Severe and near fatal asthma	Thunderstorms may trigger asthma epidemics by inducing pollen grain rupture by osmotic shock, which releases respirable allergens (0.5-2.5 μm) into the atmosphere.	Occurrence of severe asthma epidemic (during Melbourne outbreak, 9 deaths and more than 8500 emergency admissions in hospitals) during thunderstorms in the pollen season
Fan *et al.*, 2015 Sichuan Basin, China (2013 July 8-9)	Flood	Explored the contribution of anthropogenic pollution to the flood	Not considered	Model simulations of convection-permitting scale (3 km) using model WRF-Chem.	Heavy air pollution trapped in the Sichuan basin significantly enhances the rainfall intensity over the mountainous areas through "aerosol-enhanced conditional instability".

Reference Region and Period	*Extreme Weather*	*Study Linked to Air Quality*	*Health Effect*	*Indices/Factors/Methods Used*	*Results*
Wang *et al.*, 2017 USA (1990–2014)	Droughts	Examine the relationship between drought and air quality.	Not considered	Standardized precipitation evapo-transpiration index (SPEI) as Drought Index. Palmer drought severity Index (PDSI). Surface PM 2.5 and O_3.	Severe droughts were associated with mean enhancements in surface ozone and PM2.5. Estimated Increase of 1–6 per cent for ground-level O_3 and 1–16 per cent for PM2.5 in the USA by 2100 compared to the 2000s due to increasing drought alone.
Schnell and Prather, 2017 Eastern United States and Canada (1999-2013)	Heat waves	Climatology of the coincidence, overlap and lag in space and time of pollutants and maximum temperature extremes.	Not considered	Interpolation of surface hourly ozone (O_3), daily (PM2.5 abundance, and maximum temperature (TX) onto a 1×1° grid over the region	Heat waves and pollution episodes intensify under a warming climate in some regions. Warmer temperatures make O_3 pollution more severe, as the O_3 events precede temperature events. Highlights the synergistically worse situation of heat waves and air quality.
Papanastasiou *et al.* 2015 Greece, (2001–2010)	Heat waves	Link between air quality and discomfort under extreme hot weather in three Greek cities	People's thermal comfort	Common air quality index (CAQI), Thom's discomfort index (DI)	Synergetic effect between air pollution and discomfort conditions were revealed in the three cities that was more pronounced during heat wave days

Reference Region and Period	*Extreme Weather*	*Study Linked to Air Quality*	*Health Effect*	*Indices/Factors/Methods Used*	*Results*
Kimet al. 2015 South Korea (2000–2009)	Extremely hot days (daily mean temperature: > 99th per centile)	Temperature modification effect on acute mortality due to air pollution	Cardiovascular mortality.	Stratified time-series models	Highest overall risk between PM10 and non-accidental or cardiovascular mortality was observed on extremely or very hot days
Shaposhnikov *et al.* 2014 Russia (2006–2010)	Heat wave and wildfire	Day-to-day variations of deaths in Moscow, in relation to air pollution levels and temperature during the heat wave and wildfire	High excess risks for diseases of the nervous cardiovascular, respiratory, and genitourinary systems. Excess risks for cancers.	Daily average levels of PM10 and ozone. Average daily temperatures and relative humidity. Generalized linear models	Major acute effects on mortality by high temperatures and air pollution during the 2010 heat wave

different types of weather or climate extremes and air pollution leading to adverse health effects.

Several research studies have done on air quality over Sri Lanka (Seneviratne *et al.*, 2011, Seneviratne *et al.*, 2017, Ileperuma 2000). Some work has discussed the health issues related to air pollution (Ileperuma 2014, Nandasena *et al.*, 2010, Senanayake *et al.*, 2001, Dharshana and Coowanitwong 2008, Perera *et al.*, 2007), but the combined effects of meteorology and air pollution on public health have barely been examined. Moreover, detailed investigations of extreme weather or climate on air pollution and health consequences in Sri Lanka are hardly reported. PM is the major pollutant of concern in Sri Lanka due to vehicle emissions, road dust, emissions of biomass burning and sea salt. According to published past observations (Jayathilake and Jayarathne 2016), the annual average of the ambient PM10 level in the Colombo city area is in the range 72-82 μg/m^3 with a peak in 2001.

3.2 Variation of PM10 and Temperature in Sri Lanka

Figure 15.1 shows that variation of PM10 between 1998 and 2004 according to measurements at Fort, Colombo that level exceeds the maximum permissible level of PM10 in Sri Lanka, 50 μg/m^3 and WHO's latest guideline of 20μg/m^3, indicating the unhealthy air quality in the Colombo city area.

The annual concentrations of PM10 in the Colombo city area exceeded both permissible levels for Sri Lanka and WHO. PM10 are inhalable particles such as dust, dirt, soot, or smoke. Higher concentrations of PM10 in Colombo can be mainly due to combustion of fossil fuels from thermal power plants (Kelanitissa and Colombo Port Power Plants) and the motor vehicles.

The Fort air quality measurement site is next to the main railway station in Colombo. Therefore, higher PM concentrations can generate from diesel powered

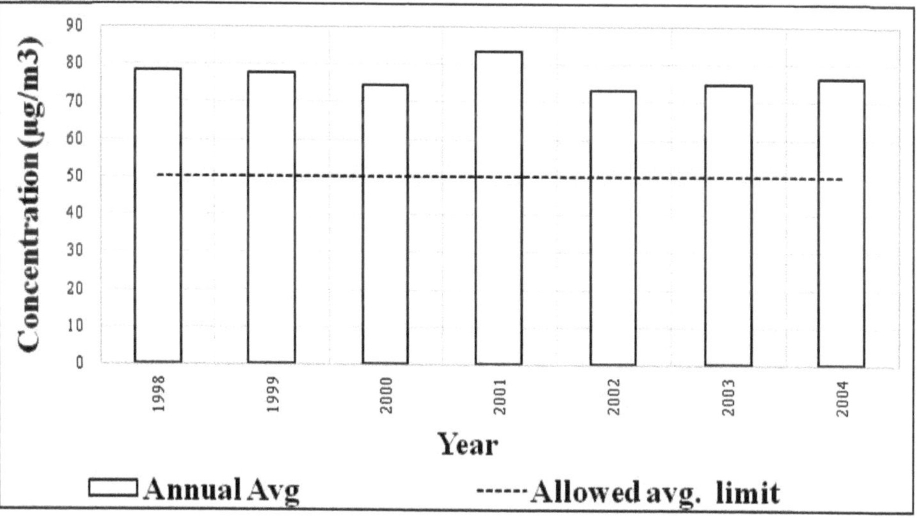

Figure 15.1: Annual Concentrations of PM10 in Colombo City Area during 1998-2004.

trains. Also, as it is a highly commercialized area, substantial particle emissions can be from automobile exhaust, particularly from old and reconditioned diesel vehicles. Further, the station is close to Colombo harbor and diesel emissions from marine vessels could be another emission source of fine particles. As an urban and industrialized area, dust, dirt, and soot can be other contributing factors of observed high PM10 concentrations. Also, the station is close to the coast and a significant percentage of sea-salt particles could be included in the observations. Particulate matter can be inhaled due to its small size and cause serious health problems, because of their ability to deeply penetrate the lungs and blood streams unfiltered. **Figure 15.2** shows extreme warm day percentages at Colombo during 1920-2004 period.

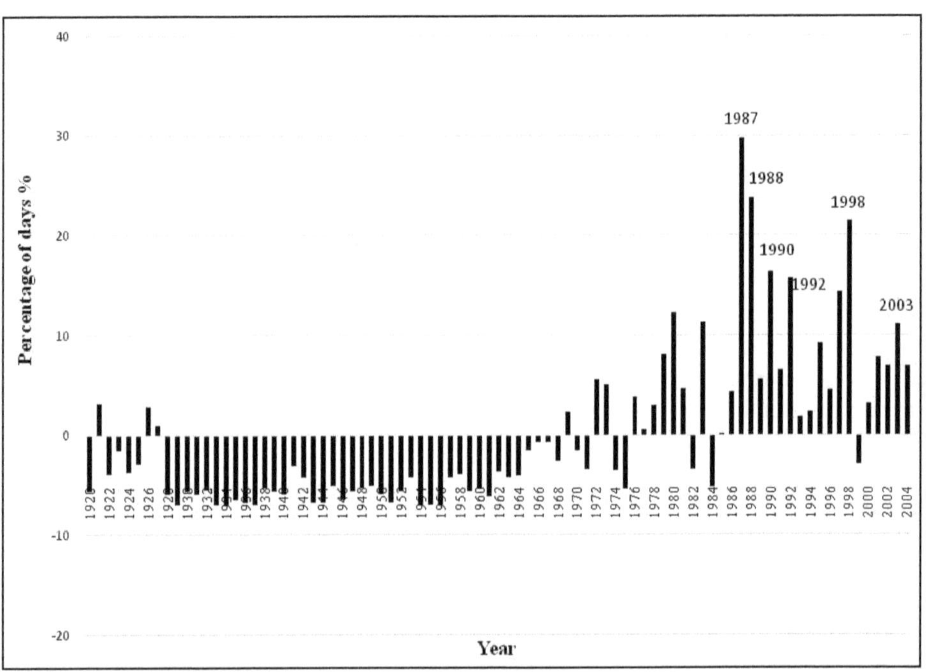

Figure 15.2: Extreme Warm Day Percentages at Colombo during period 1920-2004.

The results indicate significant increase of warm days by 10 to 20 days for Colombo. Warm days at Colombo shows negative anomalies for most years before 1970, with a negative anomaly occurring in only 1 year during the last 15 years (**Figure 15.2**). Further, the peak years (1987-88) were reported as severe drought years (Sri Lanka Disaster Knowledge Network, 2012) and 1997-98 was regarded as one of the most powerful ***El Niño***–Southern Oscillation events in recorded history (Menkeset al. 2014). The extreme warm day percentages showed a positive anomaly for all 8 stations (not shown) during the last few decades, indicating changed climate and hence weather in Sri Lanka.

All the factors observed in the above literature review and basic studies provide clues to the favored background for possible synergetic effects of heat-related weather extremes and air pollution. The situation can be worsened in future with changing climate.

4. Conclusions

Most research in the literature study provides significant evidence of air pollution-related mortality and morbidity. To determine how extreme weather and climate events affect the composition of air pollution, including formation of fine particulate matter and surface level ozone, a comprehensive literature review should be performed. Also a database of daily air pollution, meteorological variables and health indicators is needed for a considerable time period covering urban and rural areas. Air quality measurements in Sri Lanka are not adequate to cover the whole country and continuous data are not available for long measuring periods. As a solution for this deficiency in air quality measurements, the research aims to retrieve satellite-based air quality data such as *Aerosol Optical Depth* (AOD) in future work. Overall, all the evidence observed in the literature study, and basic analysis performed in this study, reveal some signs of the favored background for possible combined effects of heat-related weather extremes and poor air quality, which can cause numerous health effects.

Acknowledgements

The work is financially supported by International Science Program, Sweden and Atmospheric Physics and Lightning research group at University of Colombo (UOC), Sri Lanka. The author gratefully acknowledges Prof. I.M.K. Fernando, Department of Physics, UOC for his guidance on obtaining the air pollution data. The air pollution data used in this study are from Central Environmental Authority of Sri Lanka. The Tx90p data were taken from the global land-based climate extremes dataset produced through the coordination of the Expert Team on Climate Change Detection and Indices (ETCCDMI), Canada. Author wish to thanks all the people who have contributed all these data.

References

1. BreatheLife 2017. http://breathelife2030.org/city-data-page/?city=2400 (accessed 10 Sept. 2017)
2. Chowdhury S. and Dey S. 2016. Cause-specific premature death from ambient PM2.5 exposure in India: Estimate adjusted for baseline mortality. *Environment International* 91: 283-290
3. D'Amato G., Vitale C., D'Amato M., Cecchi L., Liccardi G., Molino A., Vatrella A., Sanduzzi A., Maesano C. and Annesi-Maesano I. 2016. Thunderstorm-related asthma: what happens and why. *Clin Exp Allergy* 46: 390-396
4. De Sario M., Katsouyanni K. and Michelozzi P. 2013. Climate change, extreme weather events, air pollution and respiratory health in Europe. *EurRespir J* 42: 826-843

5. Dharshana K.G.T. and Coowanitwong N. 2008. Ambient PM10 and respiratory illnesses in Colombo City, Sri Lanka. *Journal of Environmental Science and Health* 43: 1064–1070.

6. Fan J., Rosenfeld D., Yang Y., Zhao C., Leung L.R. and Li Z. 2015. Substantial contribution of anthropogenic air pollution to catastrophic floods in Southwest China. *Geophysical Research Letters* 42: 6066-6075.

7. Ghude S. D., Chate D.M., Jena C., Beig G., Kumar R., Barth M.C., Pfister G.G., Fadnavis S. and Pithani P. 2016. Premature mortality in India due to PM2.5 and ozone exposure. *Geophysical Research Letters* 43: 4650-4658

8. IHME 2017. http://www.healthdata.org/sri-lanka (accessed 10 Nov. 2017)

9. Ileperuma O.A. 2000. Environmental pollution in Sri Lanka: a review. *Journal of the National Science Foundation of Sri Lanka* 28: 301–325

10. Ileperuma O.A. 2014. Air pollution and increasing rate of cancer in Sri Lanka. http://www.island.lk/index.php?page_cat=article-details and page=article-details and code_title=105983 (accessed 28 Oct. 2017)

11. Intermountain Medical Center 2017. People with certain blood types are at increased risk of heart attack during periods of pollution. ScienceDaily, www.sciencedaily.com/releases/2017/11/171114155536.htm (accessed 22 Nov. 2017)

12. IPCC 2007. https://www.ipcc.ch/pdf/assessment-report/ar4/syr/ar4_syr_full_report.pdf (accessed 12 Sep. 2017)

13. IPCC 2013. https://www.ipcc.ch/pdf/assessment-eport/ar5/wg1/WG1AR5_Chapter08_FINAL.pdf (accessed 12 Sep. 2017)

14. Jain V., Dey, S and Chowdhury S. 2017. Ambient PM 2.5 exposure and premature mortality burden in the holy city Varanasi, India. *Environmental Pollution* 226: 182-189

15. Jayathilake A., Jayarathne R.N.P. 2016. Clean Air Action Plan 2025, 2016, Ministry of Mahaweli Development and Environment, Sri Lanka. available through. http://gefsgpsl.org/publication.aspx (accessed 12 Sep. 2017)

16. Kim, S.E., Lim, Y.H. and Kim, H., 2015, Temperature modifies the association between particulate air pollution and mortality: A multi-city study in South Korea;Sci Total Environ, 524–525, pp. 376–383.

17. Leelõssy Á., Molnár F., Izsák F., Havasi Á., Lagzi I. and Mészáros R. 2014. Dispersion modeling of air pollutants in the atmosphere: a review. *Open Geosciences* 6: 257-278

18. Li, Z., Niu, F., Fan, J., Liu, Y. and Rosenfeld, D., 2011.Long-term impacts of aerosols on the vertical development of clouds and precipitation. Nature Geoscience, 4(12), pp. 888–894.

19. Menkes C.E., Lengaigne M., Vialard J., Puy M., Marchesiello P., Cravatte S. and Cambon G. 2014. About the role of westerly wind events in the possible development of an El Niño in 2014. *Geophys Res Lett* 41: 6476–6483

20. Nandasena Y.L.S., Wickremasinghe A.R. and Sathiakumar N. 2010. Air Pollution and health in Sri Lanka: a review of epidemiologic studies. *BMC Public Health* 10: 300

21. Papanastasiou D.K., Melas D., and Kambezidis H.D. 2015. Air quality and thermal comfort levels under extreme hot weather. *Atmospheric Research* 152: 4–13.

22. Perera G.B.S., Emmanuel R., Nandasena Y.L.S., Premasiri H.D.S. 2007. Exposure to aerosol pollution and reported respiratory symptoms among segments of urban and rural population in Sri Lanka. *Built - Environment Sri Lanka* 7: 31–39

23. Schnell J.L. and Prather M.J. 2017. Co-occurrence of extremes in surface ozone, particulate matter, and temperature over eastern North America. *Proc Natl Acad Sci., USA* 114: 2854-2859

24. Senanayake M.P., Samarakkody R.P., Sumanasena S.P., Kudalugodaarachchi J., Jayasinhe S.R. and Hettiarachchi A.P. 2001. A relational analysis of acute wheezing and air pollution. *Sri Lanka Journal of Child Health* 30: 66–68.

25. Seneviratne M.C.S., Vajira A.W., Hadagiripathira L., Sanjeewani S., Attanayake T., Jayaratne N. and Hopke P. 2011. Characterization and source apportionment of particulate pollution in Colombo, Sri Lanka. *Atmospheric Pollution Research* 2: 207–212

26. Seneviratne S., Handagiripathira L., Sanjeevani S., Madusha D., Ariyaratna V., Waduge A., Attanayake T., Bandara D. and Hopke P.K. 2017. Identification of sources of fine particulate matter in Kandy, Sri Lanka. *Aerosol and Air Quality Research* 17: 476–484

27. Shaposhnikov D., Revich B., Bellander T., Bedada G.B., Bottai M., Kharkova T. L., Kvasha E.A., Lezina E., Lind T., Semutnikova E. and Pershagen G. 2014. Mortality related to air pollution with the Moscow heat wave and wildfire of 2010. *Epidemiology* 25: 359-364

28. Sillmann J., Kharin V.V., Zwiers F.W., Zhang X and Bronaugh D. 2013a. Climate extremes indices in the CMIP5 multi-model ensemble. Part 1: Model evaluation in the present climate. *J Geophys Res* 118: 1716-1733

29. Sillmann J., Kharin V.V., Zwiers F.W., Zhang X. and Bronaugh D. 2013b. Climate extremes indices in the CMIP5 multi-model ensemble. Part 2: Future projections. *J Geophys Res* 118: 2473–2493.

30. Sri Lanka Disaster Knowledge Network, 2012, http://www.saarc-adkn.org/countries/srilanka/disaster_profile.aspx (accessed 12 Nov. 2017)

31. SOGA 2017. https://www.stateofglobalair.org/sites/default/files/SOGA2017_report.pdf (accessed 10 Sep. 2017)

32. Wang Y., Wang M., Zhang R., Ghan S.J., Lin Y., Hu J., Pan B., Levy M., Jiang J.H. and Molina, MJ. 2014. Assessing the effects of anthropogenic aerosols on Pacific storm track using a multiscale global climate model. *Proc Natl AcadSci, USA* 111: 6894-6899

33. Wang L., Zhang N., Liu Z., Sun Y., Ji D. and Wang Y. 2014. The influence of climate factors, meteorological conditions, and boundary-layer structure on severe haze pollution in the Beijing-Tianjin-Hebei region during January 2013. *Advances in Meteorology* 14: 1-14

34. Wang Y., Xie Y., Dong W., Ming Y., Wang J. and Shen L. 2017. Adverse effects of increasing drought on air quality via natural processes. *AtmosChem Phys* 17: 12827–12843

35. WHO 2016. http://www.who.int/mediacentre/news/releases/2016/air-pollution-estimates/en/ (accessed 12 Sep. 2017)

36. WHO 2005. http://apps.who.int/iris/bitstream/10665/69477/1/WHO_SDE_PHE_OEH_06.02_eng.pdf (accessed 11 Sep. 2017)

37. Zeng S. and Zhang Y. 2017. The effect of meteorological elements on continuing heavy air pollution, a case study in the Chengdu area during the 2014 spring festival. *Atmosphere* 8: 85-94

38. Zhang B., Wang H., Park Y., T.W and Deng Y. 2017. Quantifying the relationship between extreme air pollution events and extreme weather events. *Atmos Res* 188: 64–79.

Chapter 16

Assessment of Desertification in Iran by the ANN and IMDPA Model

A.R.N. Namaghi[1]* *and Z. Golizadeh*[2]

[1]*Department of Natural Resources, Khorasan-e-Razavi Agricultural and Natural Resources Research and Education Center, Mashhad, Iran*

[2]*CEO of Asia Ecosystem Institute, Iran*

**E-mail: arnamaghi@gmail.com*

ABSTRACT

Desertification and climate change are two important factors in the arid ecosystems. More than a billion people in the world are threatening by Desertification as an Extreme Natural Event. Iranian Desertification Model Potential Assessment (IMDPA) is a model for studies and to assess desertification in Iran, in this model, a lot of indicators and criteria were considered. Indicators of Climate, Geology, Geomorphology, Soil, Vegetation Cover, Agriculture, Water and Erosion are the most important environmental factors for desertification assessment in Iran. Artificial Neural Networks (ANN) the idea is to process information that inspired by biological nervous system such as the brain to process information. Environmental indicators for assessing the severity of desertification have more different criteria with unknown different weights. Result of this research show Neural Networks and Genetic Algorithms can be used to optimize environmental indicators and exact weight of it in this model.

Keywords: *Climate change, Arid ecosystem, Environmental indicator, Artificial neural network.*

1. Introduction

Desertification and land degradation, are caused leading to instability and the formation of desert in arid and semi-arid lands. Many environmental and non-environmental factors are involved in this process. There are a lot of evaluation indicators of desertification in different models and different levels. The important model of desertification evaluation include: **IMDPA** (*Iranian Model of Desertification*

Potential Assessment), **FAO-UNEP** (*Food and Agriculture Organization-United Nations Environment Programme*), **LADA** (*Land Degradation Assessment in Dry Lands*), **MEDALUS** (*Mediterranean Desertification and Land Use*), **GLASOD** (*Global Assessment of Human-induced Soil Degradation*) and so forth which can be in the form of **DPSIR** (Driving Forces, Pressure, State, Impact and Responses) studied. In the land degradation (LD) or reduced production, as well as the concepts and indicators of desertification would be used. Indicators of Climate include: annual rainfall, drought, indicator of geology and geomorphology include: physiographic, rocks sensitivity, type of operation in the work unit, indicator of soil include: the percentage of gravel, soil depth, soil texture,electric conductivity, indicator of vegetation cover include: renewal index cover, vegetation condition, usage of vegetation cover, indicator of agricultural include: cropping patterns, crop yields, use of inputs and machinery, indicator of water include: electrical conductivity, sodium adsorption ratio, groundwater level. indicator of erosion include: erosional facial appearance parameters, density non-life coverage, vegetation, number of days with dust storm index, type and density of water erosion, the type of land use, density of vegetation, from The most important environmental factors in the assessment of desertification to Iran. ***Artificial Neural Networks*** (ANN), concept for the processing of information that is inspired by biological nervous systems, such as the brain, to process information.

The neural network formed from number of node or single cell or neuron, which input set, is connected to the output.The ability of neural networks is to obtain meaning from complicated or imprecise data.The benefits of adaptive neural network learning, self-organizing, performing calculations in parallel, without interrupting fault tolerance and disadvantages of the absence of specific instructions to design for a specific application and the size of the training set is dependent on the accuracy of results.

The main objective of this research is to determine the environmental factors for the instability of the land in the study area, with environmental index of desertification and the use of artificial neural networks and genetic algorithms, and to determine the weights of each environmental indicator of desertification, optimization of environmental indicator of desertification potential by genetic algorithms.

2. Methodology or Information Analysis

The study area consists of a watershed in Khorasan Razavi province in Torbat-e-Jam City. Basic studies of natural resources, including physiographic, soil, vegetation cover, geology and geomorphology, hydrology and water resources and environmental criteria to determine a basic map was IMDPA model. Using the evaluation plan was prepared. Using artificial neural networks and genetic algorithm with the help of software NeuroSolutions ver. 5.07, according to the following chart to determine the weight of each of the proposed environmental foundation with the help of input parameters and plans of potential desertification and land degradation measure and weigh each of the criteria defined in the model. Saleh Abad Torbat-e

Jam in the study area as a hydrological catchment areas with an area of 6745 ha and is approximately 47 km from the North East and 44 km South-West of Saleh Abad Torbat city is located. This area is located near the village of BaniTak exit point. The average, minimum and maximum height of area are 1290, 1916 and 2130 meters, respectively and the weighted average slope area of about 30.7 per cent.

The status of desertification is the current condition of study area. Criteria and indicators of effective or the low intensity of desertification condition determined according to studies and basic maps.The results of the assessment of desertification,Artificial Neural Network algorithm and based on numerical criteria are evaluated. Artificial Neural Network Based on the weighting of each of the input data, the ability to provide the value of each input variable output in the selection of the final model compared with the control data. ANN structures MLP (Multilayer Perception Network), GFF (General Feed Forward Network), and RBF (Radial Basis Function Network) and Education is analyzed by algorithms. All the factors contributing to desertification process in ArcGIS and using the overlaying layers combined (**Figure 16.1**).

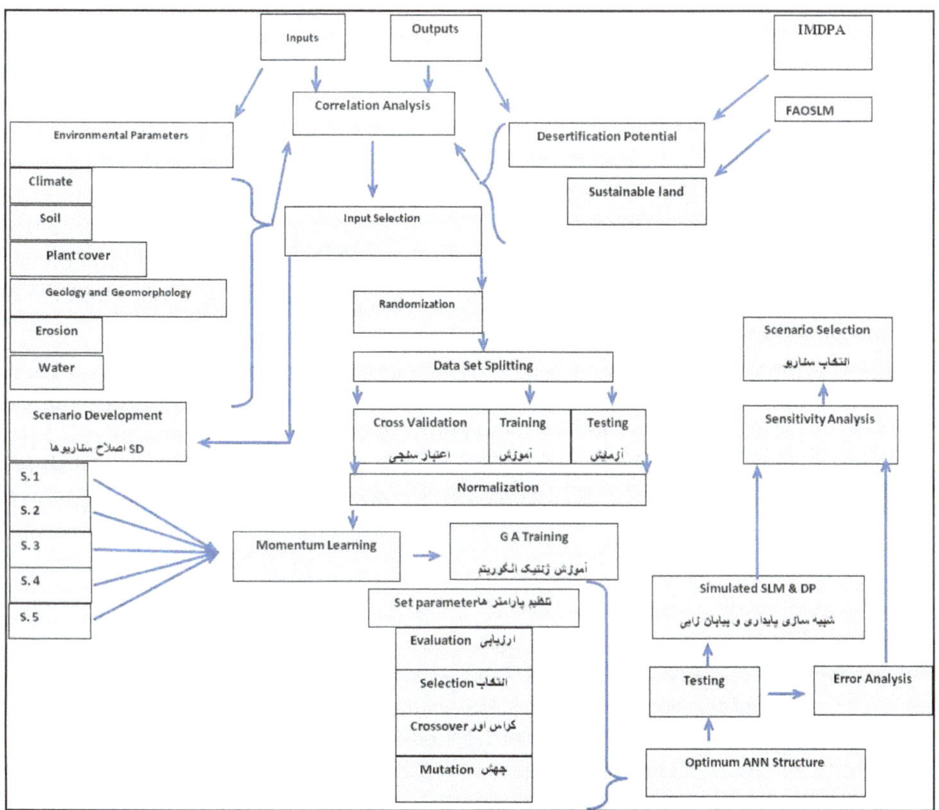

Figure 16.1: Diagrams of Research Steps Used in the Chapter.

3. Results and Discussion

Annual precipitation, is 270 mm and mean annual temperature of the region area is 11.6 degrees Celsius.The highest monthly rainfall of area is 59 mm in March and in August; rainfall is less than 1 mm. The annual rainfall between 210 mm in low area and 340 mm in high points is variable. The maximum amount of water from melting snow in February is 30 mm. The climate of area is Semi-arid or cold and semi-arid estimated. Geological formations that outcrop in the field are the second and third geology period. These units, from old to new, including Chile Formation, Upper Triassic Sandstone, that their name known as valley (TRm). That is the basic unit of shale and sandstone lithology, and layers of sandstones are more abundant than in Chile. Within this formation, there are other lithological units, in terms of lithology is mostly sandstone, there is Chile layers in between layers.This unit is part of the Kashafrud formation are considered. Kashafrud and Miankuhare the most of lithological formations comprises in the study area. In terms of geomorphology; in general, the field of study divided two units; rugged mountains and dunes. Greater extent contributed to the dunes.In the mountain unit defined 5 types of lithology and 8 geomorphologic facies. The study area consists of 4 types of physiographic; Mountains, hills, plateaus and alluvial plains of the river. Important limiting factors in mountain brigades rock outcrops, slope ups and downs and a lot of vegetation or soil and water erosion is poverty.Among the important operations development in this area are strengthens the vegetation, control grazing, soil erosion control and mechanical watershed management operation. Part of the land has the ability as a protected forest, the rest of the land has a low potential for pasture.Conditions hills, like the mountain and only parts of the territory1-3-2, has low potential for agriculture, in the alluvial plains and river, erosion is one of the limiting factors and the land has the ability to use their agriculture. Soils of the field of study placed in two categories: ***Entisols*** and ***Inceptisols***. The study area has 5 vegetation cover type, with the following characteristics:

***Type I*:** This type with an area of 689.3 ha (9 per cent surface area) had a dominant species, *Artmisia aucheri* and *Asteragalus gummiferous* with canopy cover 37.4 per cent, bare soil 22.6 per cent, gravel 37.5 per cent and litter 2.5 per cent, with poor ecological state and the negative trends range condition.

***Type II*:** This type with an area of 2198.9 ha (28.6 per cent of total area) had a dominant species, *Perovkia abrotanoides* and *Asteragalus gummiferous*, vegetation cover 39.4 per cent, bare soil with23.4 per cent, gravel 35.9 per cent, litter1.3 per cent, negative trend range condition and poor ecological status.

***Type III*:** This type with an area of 1163.3 ha (15.2 per cent of total area) are the dominant species *Ephedra major*, *Perovskia abrotanoides*, *Pistacia vera* with vegetation cover 31.2 per cent, bare soil with 12.1 per cent, gravel 54.6 per cent, litter 1.2 per cent, negative trend range condition and poor ecological status.

***Type IV*:** This type with an area of 1697.6 ha (22.1 per cent of total area) are the dominant species *Artemisia herbalba*, *Poa bulbosa* with vegetation cover

51.9 per cent, bare soil with 31 per cent, gravel 15.9 per cent, litter 1.2 per cent, negative trend range condition and poor ecological status.

***Type V*:** This type with an area of 132 ha (1.7 per cent of total area) are the dominant species *Peganum harmala* with vegetation cover 38 per cent, bare soil with 37 per cent, gravel 25 per cent, litter 0 per cent, negative trend range condition and poor ecological status.

According to data of the area, village population is 1036 people and 198 households. Household size of 2.5 people in the area, the relative density is 30 people per square kilometer.

Livestock in the study area in terms of revenue and the main part of the income to be includedas 66.6 per cent of GDP to account for. Deposition rates started from 58.1 tons per hectare per year in sub basin B′5 up to 97.3 tons per hectare in sub basin B3. B1 sub basin also has a higher deposition than the other sub basin (see **Table 16.1 and Figure 16.2**).

Table 16.1: Table of Study Area's Sub-basins Characteristics

Sl.No.	*Code of Sub-basin*	*Type*	*Perimeter (Km)*	*Area (Km²)*
1	B1	Hydrologic	16.03	10.48
2	B2	Hydrologic	7.87	2.96
3	B3	Hydrologic	14.29	8.19
4	B4	Hydrologic	10.61	3.56
5	B5	Hydrologic	13.25	5.09
6	B6	Hydrologic	17.07	11.79
7	B7	Hydrologic	8.49	3.48
8	B'1	Non hydrologic	3.05	0.39
9	B'2	Non hydrologic	2.39	0.28
10	B'3	Non hydrologic	11.6	5.40
11	B'4	Non hydrologic	14.30	6.06
12	B'5	Non hydrologic	3.67	0.57
13	B01=B5+B6+B'1	Hydrologic and non-hydrologic	20.24	17.29
14	B02=B01+B4+B7+B'2	Hydrologic and non-hydrologic	23.01	24.59
15	B03=B02+B3+B'3	Hydrologic and non-hydrologic	31.31	38.18
16	B04=B03+B2+B'4	Hydrologic and non-hydrologic	32.46	47.20
17	BT=B04+B1+B'5	Hydrologic and non-hydrologic	39.44	58.25
18	N1	Hydrologic	17.29	8.89
19	N2	Hydrologic	16.68	9.63
20	N'1	Non hydrologic	0.83	0.02
21	NT=N1+N2+N'1	Hydrologic and non hydrologic	19.79	18.51

In this research, desertification intensity of 3 study area was evaluated using IMDPA, the newest method of assessment of desertification potential in arid and

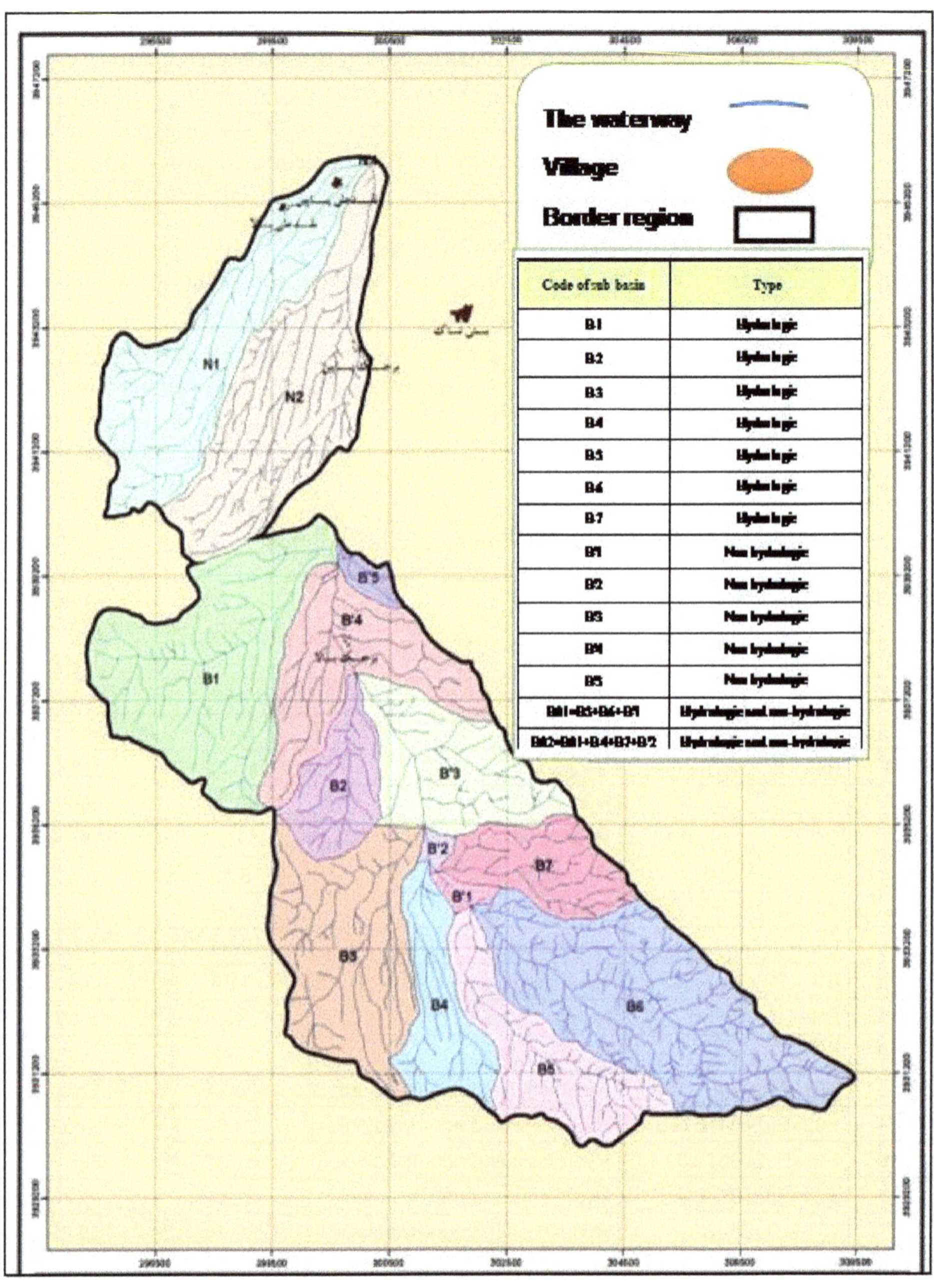

Figure 16.2: Map of Sub-basin of Study Area.

semi-arid regions of Iran. For this purpose 4 criteria selected including vegetation cover, soil, and climate and wind erosion (**Figure 16.3**). Each criterion was assessed based on the selected indices which result in qualitative mapping of each criterion

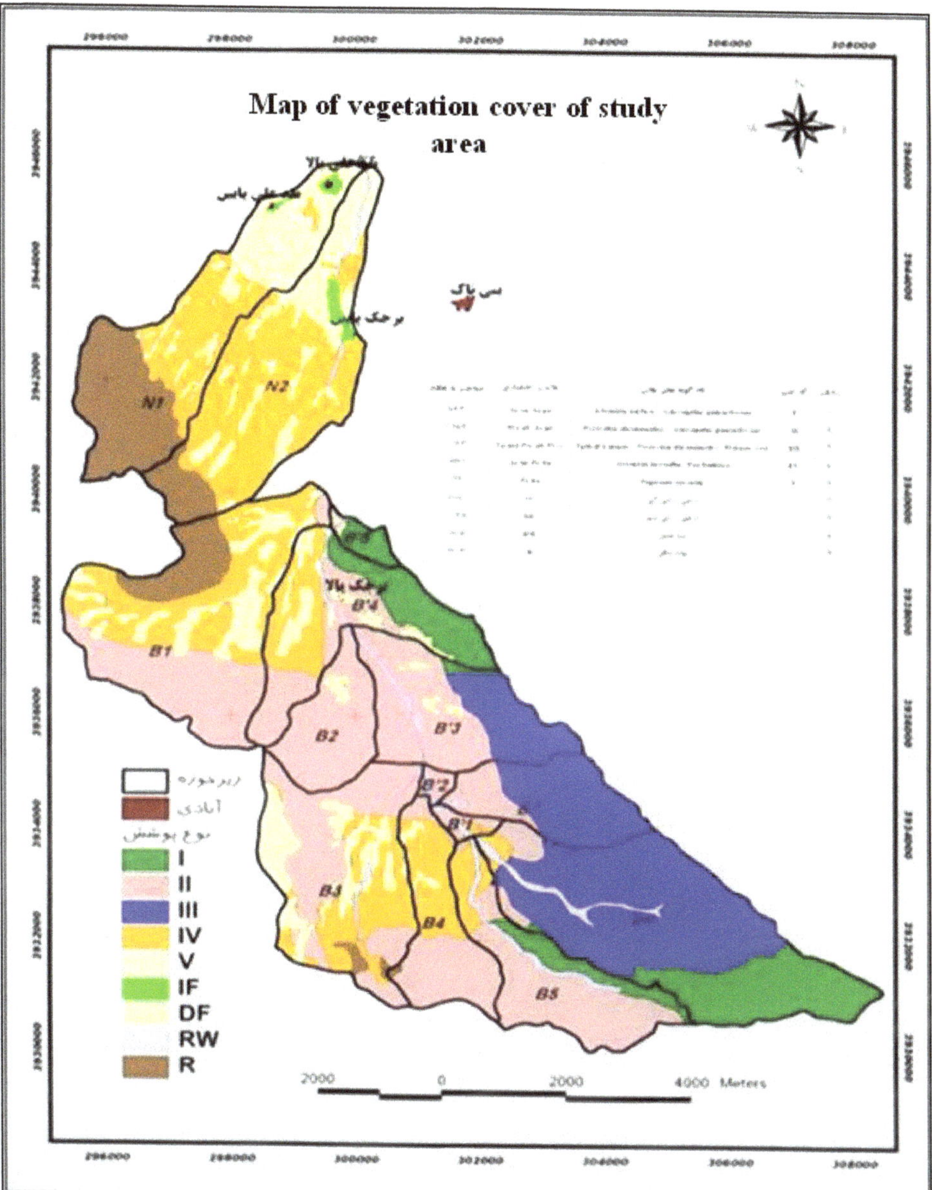

Figure 16.3: Map of Vegetation Cover Typology of the Study Area with Sub-basin and Villages.

cased on geometric average of the indices. Then, sensitive map of region was extracted using geometric average of all criteria. Thematic databases, with a 1:50000 scale resolution, were integrated and elaborated in a GIS based on arc/info8, arc view 3.2 and especially ILWIS. By laminate of thematic databases layers and using this formula DM= (SI×WEI× VI×CLI) 1/4 to calculate the geometric mean criteria.

Then desertification intensity map was obtained with analysis IMDPA Model and ANN. The result showed that 51.09 per cent of study area was found to be in medium and about 45.09 per cent in high class of desertification. And while 3.82 per cent of the region in clouding residential water tanks (CHAH NIME) there aren't in any classification. Analysis of desertification criteria in study area region showed that

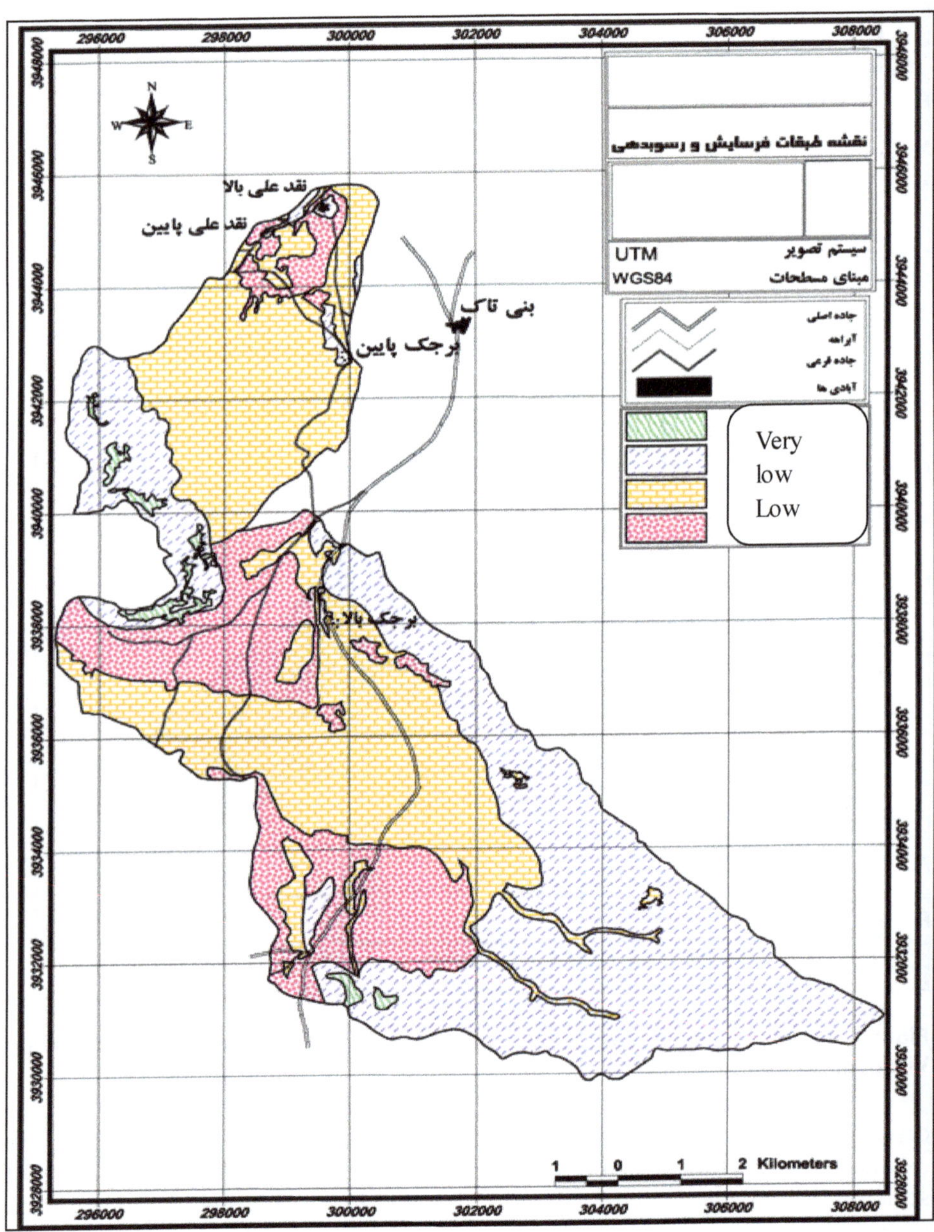

Figure 16.4: Map of Study Area's Erosion and Sediment Classes.

among study criteria, wind erosion criteria is a major problem with a geometric average of 1.67 which shows high class while soil with a weighted average of 1.34 has lowest effect in desertification (**Figure 16.4**). Also vegetation cover and climate criteria with a geometric average of 1.51 and 1.57 show middle class desertification.

4. Conclusions

By comparing the prepared desertification maps and assessing sustainable land management, the most important known environmental index in each criterion include Drought index in the climate criterion, vegetation cover index in vegetation cover, land use type index in the measure of water erosion and vegetation index in the wind erosion criterion in the study area. Among the criteria for assessment, vegetation and erosion criteria, show the highest weight for the region's instability and the severity of desertification. Due to the importance of this problem, various methods for estimating or evaluating desertification have been presented and criticized so far. However, despite the progress made, there are many problems with the use of desertification assessment methods, including the fact that these methods are still largely based on judgment and expert opinion that, with disagreement with the public opinion Relations with degraded lands have even led the national level.

References

1. Catherine *et al.*, 2004. Zoning for stability in a dry climate with emphasis on groundwater resources management, *Water Resources Research Center*, University of Arizona, AZ, USA.
2. Steer M. *et al.*, 2012. To evaluate the sustainability of small farms and natural resources management system (MESMIS). *Institute of Geography*, National University of Mexico, Mexico.
3. Apink F. *et al.*, 2012. Land management and ecosystems. *Center for Environmental Research* in Germany, Germany.
4. Martin *et al.*, 2001. Integrating land use and environmental factors for the sustainable management of groundwater resources. *Water services committee hydrology occupied* Palestine, Israel.
5. Aschavylych J. *et al.*, 2010. Experience in monitoring and evaluation of land management. *Centre for Development and Environment*, University of Bern, Germany.
6. Saeed A.M.A. 2011. The nature and causes of land degradation and desertification in Libya requires sustainable management of land. *Geographical Program Faculty of Education*, University of Science, Malaysia.
7. Anonymous 2007. The use of artificial neural network (ANN) for mapping soil stability.
8. Rychartr B.D. *et al.*, 2003. From the perspective of ecological sustainable water resources management. *Nature Protection Research Center* in Virginia, USA

9. Reed M.S. *et al.*, 2002. Selection criteria, determine the range in the Kalahari for the sustainable management of natural resources, *Development and Environment*, University of Leeds, Leeds, UK
10. Samuel B.A. *et al.*, 2012. Criteria for sustainable management of land and its fuzzy in Nigeria. *Faculty of Extension and Agricultural Economics*, University Northwest, USA
11. Wieland A. *et al.*, 2010. Criteria for sustainable management of land use in Chile, *Institute of Geography*. University of Leipzig, Germany
12. Vlad V *et al.*, 2002. General methods for assessing the sustainability of land use and basic criteria for continued agricultural land use. *Soil science and agricultural chemistry research*
13. Reed M.S. *et al.*, 2011. Monitoring and evaluation of cross- scale of Tkhyryb land and sustainable management of land, as the methodology for the management of environmental sustainability and the *Centre for Environmental Planning and Management*. Faculty of Earth Sciences at Aberdeen University, England
14. Smith A.J. *et al.*, 1995. A structure for evaluating the sustainable management of land, *Research Center for Agriculture and Food Industry of Canada*, Ottawa, Canada
15. Yamamoto Y.Q. *et al.*, 1996. Application of neural network and GIS in land evaluation models, *Japan's National Research Institute of Ranch*.
16. Lefevre A.D.B. *et al.*, 2000. Land sustainability criteria on the basis of investigations carried out on farms in Thailand, Indonesia, Vietnam, *Thailand International Institute of Soil Research*, Thailand
17. Cassius J. *et al.*, 2013. And threshold criteria for evaluating the stability of the country, the International *Center for Soil Research and Management*, Thailand
18. Rasim M. *et al.*, 2010. Dynamic simulation model of desertification in Egypt. *Faculty of Computer and Information*, Cairo University, Egypt
19. Hyvrny H. *et al.*, 2000. To assess the sustainable management of *land and environment development center* in Switzerland.
20. Hldn A. *et al.*, 2005. The perception of desertification, the *Sixth Committee of the Europe program*
21. Anonymous 2010. *The sustainable management of land* (Monitoring Guide)

Section IV

Landslides

Chapter 17

Landslide Risk Assessment and Real Time Monitoring for Minimizing the Impact of Rainfall-Induced Landslides in Indian Himalayas

D.P. Kanungo

CSIR-Central Building Research Institute,
Geotechnical Engineering Group, Roorkee – 247 667, Uttarakhand, India
****E-mail: debi.kanungo@gmail.com***

ABSTRACT

Landslides affect about 15 per cent of Indian landmass aggregating to about 0.49 million square kilometers of area. Landslides cause destruction and losses to lives and properties each year in 22 States and Union Territory of Puducherry. During recent years, the impact of landslide disasters has compounded due to extreme natural events, urbanization in hill areas and non adherence of land use policies. Thus the severity and impact of disasters are increasing both in terms of magnitude and frequency of occurrences affecting the GDP and development of region. Therefore, landslide risk assessment and real time monitoring is the need of the hour for minimizing the impacts of extreme natural event such as rainfall in Indian Himalayas. In light of this, the present paper discusses the following aspects in Indian context: Combined neural and fuzzy technique for landslide susceptibility zonation, Rainfall thresholds for prediction of shallow landslides; Ground based wireless instrumentation and real time monitoring of a landslide; Quantitative estimation of physical vulnerability of buildings, exposed to landslides.

Keywords: *Landslide susceptibility, Rainfall threshold, Real time monitoring, Physical vulnerability.*

1. Introduction

The framework of landslide risk assessment and real time monitoring encompasses identification of potential landslide slopes and associated precursor phenomena, reliable landslide hazard and risk assessment, development of reliable and durable slope monitoring system and slope conservation and disaster mitigation measures. Landslide Susceptibility Zonation (LSZ) is necessary for planning future developmental activities in hilly areas. Many techniques in this direction have been developed in last few decades all over the globe. In this paper, a novel technique based on combined neural and fuzzy set theory concepts for preparation of LSZ maps is discussed. Majority of landslides in the Indian sub-continent are triggered by rainfall. Several attempts have been made to establish rainfall thresholds on global, regional and local scales for the occurrence of landslides. In this paper, an attempt is made towards deriving local rainfall thresholds for landslides in terms of intensity duration and antecedent rainfall based on daily rainfall data in parts of the Garhwal Himalayas, India. The installation of a real-time monitoring system is often a cost effective risk mitigation measure. This paper discusses the establishment of a 'Landslide Observatory' with wireless instrumentation for real time monitoring of ground deformation and hydrologic parameters at Pakhi Landslide in Garhwal Himalayas, India. Due to population growth and unplanned urbanization, people tend to move towards unstable slopes to construct buildings and thereby making structures vulnerable due to landslides. This paper further presents a quantitative approach to assess scenario based physical vulnerability of buildings based on resistance ability of the buildings and landslide intensity.

2. Methodology Overview

2.1. Combined Neural and Fuzzy Technique for Landslide Susceptibility Zonation

The genesis of the combined neural and fuzzy approach discussed here for LSZ mapping lies in the objective determination of weights and ratings to be assigned to various causative factors. The weights of thematic data layer pertaining to each causative factor were determined *via* Artificial Neural Network (ANN) connection weight analysis method, whereas ratings of the categories of each thematic layer were determined through cosine amplitude fuzzy similarity method. The rating value for a category can be defined as the ratio of total number of landslide pixels in the category to the square root of the multiplication of total number of pixels in that category and the total number of landslide pixels in the area. Values close to 0 indicate dissimilarity whereas values close to 1 indicate the similarity between two datasets. A feed forward multi-layer ANN is adopted for weight determination of causative factors, which comprises of one input layer, one output layer, and two hidden layers in between input and output layers. In the present case, six thematic data layers (causative factors) correspond to the six neurons in the input layer and a single neuron in the output layer represents the desired output, indicating an existing landslide or no-landslide locations. The neurons in a layer are connected to the neurons of the next successive layer and each connection carries a weight. In this study, 100 neural networks were designed, trained and tested with three independent datasets, one each for training, testing and verification of the neural network consisting of landslide pixels and no-landslide pixels. The adjusted weight matrices (one each for input-hidden, hidden-hidden and hidden-output layers) were sequentially multiplied to generate the final weight matrix to indicate weight assigned to each factor. After taking the absolute values of these weights, these were ranked in descending order. Considering all the 100 networks, the rank of a factor was decided based on the rank observed by maximum number of networks (majority rule). Subsequently, the normalized average of the weights of these networks at a scale of 0 to 10 for a particular factor was calculated and assigned as the weight of that factor for preparation of LSZ map. The weights assigned to factors and the ratings assigned to categories were integrated in GIS to produce LSZ map. The final LSZ map showed landslide areas categorized into VLS, LS, MS, HS and VHS zones. The details of the methodology, analysis and results are discussed in Kanungo *et al.*, 2006, 2009 and Gupta *et al.*, 2008.

2.2. Rainfall Thresholds for Prediction of Shallow Landslides

An attempt towards deriving local rainfall thresholds for landslides based on daily rainfall data in and around Chamoli-Joshimath region of the Garhwal Himalayas, India is made. To understand the relationship between the rainfall pattern and the landslide occurrences, threshold analysis based on rainfall intensity duration and antecedent rainfall concepts has been tried out. Rainfall events pertaining to 81 landslides were analyzed to yield empirical intensity duration threshold for landslide occurrences. The rainfall threshold relationship fitted to lower boundary of landslide triggering rainfall events is $I = 1.82\ D^{-0.23}$ (I = rainfall intensity in mm per hour and D=duration in hours). It is revealed that for rainfall events of shorter duration (≤ 24h) with a rainfall intensity of 0.87 mm/h, the risk of landslide occurrence is expected to be high. Further, antecedent rainfall is responsible for predisposing the hill slopes to failure as it controls the soil moisture level in the process of slow saturation of soil layer and also influences the ground water level. A simple way to utilize these accounts for establishing a threshold based on antecedent rainfall prior to the landslides. Monsoon rainfalls occur with interruptions and are characterized by low intensity and long duration with occasional cloud bursts. Therefore, the role of antecedent rainfall in initiating or reactivating landslides has been analyzed by considering daily rainfall at failure and different period (3, 7, 10, 15, 20 and 30 days) cumulative rainfall prior to failure for 128 landslides. It is observed that a minimum 10-days antecedent rainfall of 55 mm and a 20-days rainfall of 185 mm are required for the initiation of landslides. The details of the methodology, analysis and results are discussed in Kanungo and Sharma (2014).

2.3. Ground Based Wireless Instrumentation and Real Time Monitoring of a Landslide

Landslide instrumentation and real-time monitoring can provide insight into the understanding of the dynamics of landslide movement. The installation of a real-time monitoring system is often a cost-effective risk mitigation measure. A Landslide Observatory with wireless instrumentation for real time monitoring of Pakhi Landslide in Garhwal Himalayas, India is established. With the combination of surface and sub-surface mounted monitoring sensors, actual displacements/

movements in the landslide area and weather conditions that affect the sliding activity have been targeted. Surface sensors include Wire-Line Extensometers (WLE) and Automatic Weather Station (AWS). Three numbers of extensometers have been installed across the radial tension cracks to monitor the detachment of the downhill block from the uphill block. Sub-surface sensors include biaxial In-Place Inclinometers (IPI) installed at different depths within a particular material and in the interface zones down the bore hole (BH) to measure the sub-surface displacements and Vibrating Wire Piezometers (VWP) in the bore holes to measure the variation in pore water pressure. All these surface and sub-surface sensors except AWS are connected to the nodes on the surface and communicating wirelessly with the gateway placed in field control station. AWS is connected to Data Acquisition System (DAS) also placed in field control station. The data from the field control station are being transferred on real-time to the control computer at CSIR-CBRI, Roorkee (India) through web server using ARGUS monitoring software for analysis and visualization. The real time data is being monitored to establish warning thresholds. The detail of this work is discussed in Kanungo *et al.*, 2017.

2.4. Quantitative Estimation of Physical Vulnerability of Buildings Exposed to Landslides

Landslides are also responsible for significant damage to critical infrastructure and buildings *etc.* A quantitative approach to assess scenario based physical vulnerability of buildings based on resistance ability of the buildings and landslide intensity is proposed. Physical Vulnerability (PV) is defined as a function of landslide intensity (I) associated with buildings at risk and resistance ability (R) of the buildings to withstand the threat (Li *et al.*, 2010). The modified equation is given as follows -

$$PV = f(I, R) = \{(2(I_m^{\,2})/(R_{bld}^{\,2}) \qquad I_m/R_{bld} \leq 0.5 \tag{1}$$

Where I is the landslide intensity, R_{bld} is the resistance of the building at risk and I_m is the landslide intensity of different degree m (m varies from 1 to 4 with low, moderate, high and very high intensity). The value of Physical Vulnerability (PV) is expressed non-dimensionally between 0 and 1. Both the intensity and resistance are also expressed in dimensionless terms between 0 and 1. During landslides, structures may experience damage or cracking/collapse or complete destruction. Thus, to know the status of buildings at any landslide intensity and resistance factor of the building, vulnerability has been classified into three categories as follows:

(a) **Class I:** When $PV \leq 0.5$, then structures will experience only certain amount of damage;

(b) **Class II:** When $0.5 < PV < 1$, then structures might undergo partial cracking in the walls or collapse of some structural element;

(c) **Class III:** When $PV = 1$, then structures will deform completely.

In this approach, the landslide intensity (I) is defined as a function of landslide volume (í) and landslide expected velocity (s) (Cardinali *et al.*, 2002).

$$I_m = f(v, s) \tag{2}$$

Herein, the landslide intensity is estimated by taking into account the landslide velocity scale (Cruden and Varnes, 1996) ranging from very slow (5×10^{-10} m/s) to extremely rapid (5 m/s) and the landslide volume classes (Sarkar *et al.*, 2015) ranging from low (<1000 m^3) to very high(>1000000 m^3). Consequently, the degree of landslide intensity for different combinations of landslide velocity and volume can be represented in the form of a landslide intensity matrix. The landslide intensity matrix is classified into low, moderate, high and very high classes and further assigned values on a scale of 0 to 1.

In the present study, resistance factors for five primary structural parameters namely structural typology (ξ_{sty}), state of maintenance (ξ_{smn}), construction quality (ξ_{scq}), number of storey (ξ_{sst}) and building site location (ξ_{ssl}) have been considered. The resistance of structure is estimated as a function of different resistance factors (Li *et al.*, 2010):

$$R = \left(\prod_{i=1}^{n_s} \xi_i \right)^{1/n_s} \tag{3}$$

Where ξ_i is the ith of $n_s \geq 1$ resistance factors (each expressed in non-dimensional terms) contributing to the definition of different categories. Hence, the resistance of buildings (R_{bld}) based on these five parameters can be quantified as follows:

$$R_{bld} = (\xi_{sty}, \xi_{smn}, \xi_{scq}, \xi_{sst}, \xi_{ssl})^{1/3} \tag{4}$$

The resistance factors for all the categories within each structural parameter are assigned based on their probable resistance in case of exposure to a landslide event on a scale of 0 (least resistant) to 1 (most resistant).

3. Discussion and Summary

The combined neural and fuzzy approach is proved to be one of the best objective approaches for LSZ mapping. In India, most of the LSZ exercises have been carried out on 1:50000 scale which has very limited usage in the actual ground for planning developmental activities. Hence, there is a need for large scale (1:5000 or less) LSZ mapping for its practical application for societal benefit. This can only be achieved with the use of high resolution remote sensing data and field based ground information.

Though the rainfall threshold based landslide early warning is at a very pre-mature stage in Indian scenario, this attempt has shown very exciting results in terms of intensity-duration and antecedent thresholds for landslide occurrences. The antecedent rainfall threshold concept can also differentiate landslides occurring due to cloud bursts from those occurring due to low intense high duration rainfall. The rainfall threshold based models can be improved upon with hourly rainfall data instead of daily rainfall. The threshold models should be validated with data of later years and can be refined simultaneously to achieve more accurate temporal predictions of landslides. These thresholds can be used in forewarning of landslides to guide the traffic and provide safety passage to the tourists travelling along the pilgrim routes during monsoon seasons. Also, it can be used for forewarning the landslide events for different catchment areas having cluster of villages in hilly region to save life and property. On establishing a rainfall threshold for a particular terrain having a specific lithotectonic and geo-morphological setup, a number of AWS units in the region can be wirelessly networked at a central station and an early warning can be issued. In a contrary, extensive conventional and cost intensive wireless instrumentation and real time monitoring of landslides for establishing early warning can be adopted for perennially active landslides in strategic areas

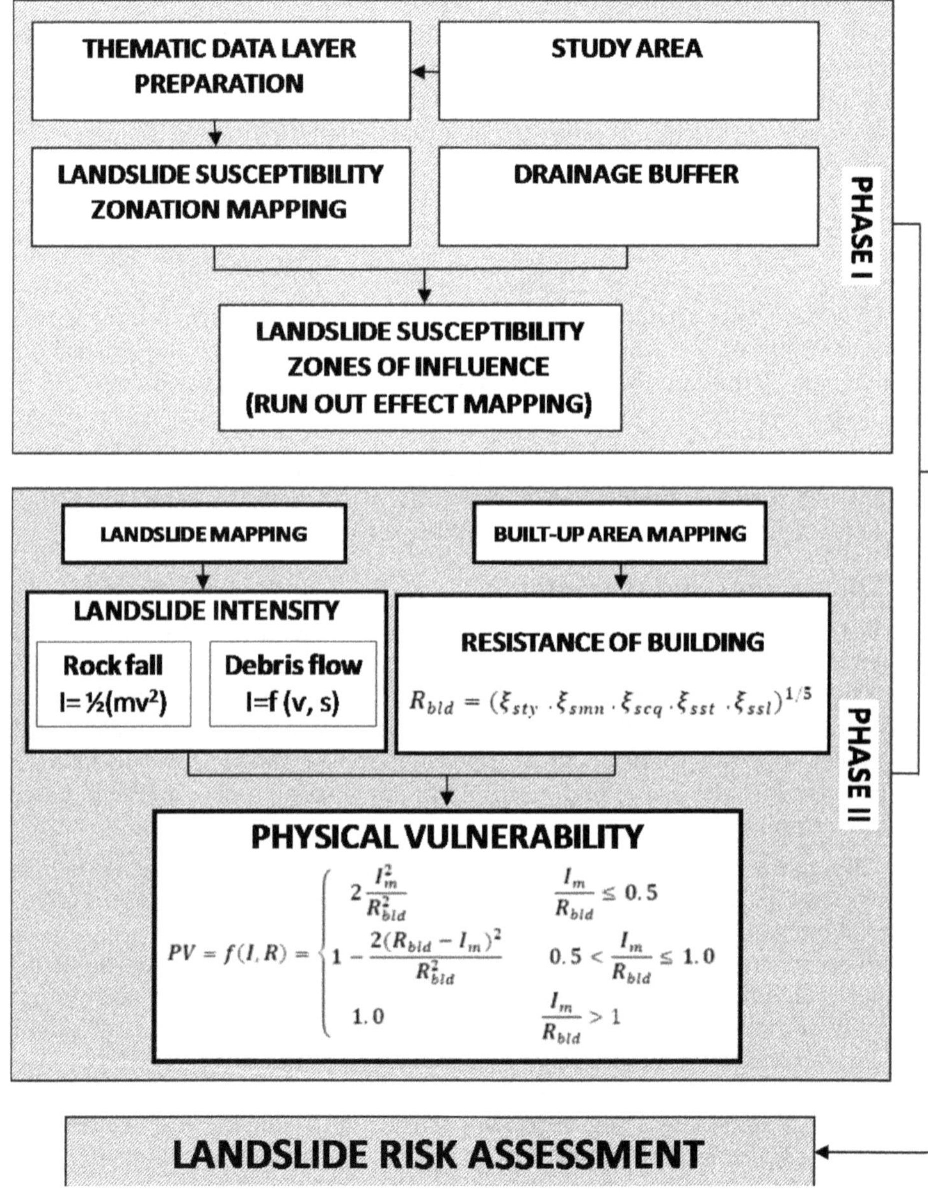

Figure 17.1: Executable Framework for Landslide Risk Assessment.

only. This approach has a very promising future if the costly conventional sensors can be replaced by the low cost MEMS based sensors.

Structures in the mountainous regions are vulnerable to landslide hazard. The proposed methodology intends for the assessment of physical vulnerability of the buildings in hilly terrain susceptible to different degrees of landslide occurrences, particularly debris slides. One of the major limitations of the proposed approach is that this may not be applicable to landslides of rock fall/topple type which needs to be developed. Further, there is a scope of improvement in this vulnerability approach by incorporating more and more essential structural parameters like foundation type, roof structure type, re-entrant corners, geometry of structure, regularity/irregularity in elevation, presence of large windows built towards the mountains slopes *etc.*

With such contributions towards landslide disaster mitigation in Indian Himalayas, an executable framework for landslide risk assessment (**Figure 17.1**) can be drawn involving hazard zonation for identification of landslide source areas, influence zone demarcation through run-out-effect mapping, mapping of risk elements and their vulnerability.

Further, rainfall threshold based models and MEMS based ground sensor network for real time monitoring of landslides can be worked upon to develop a reliable and cost effective landslide early warning system for Indian Himalayas.

Acknowledgements

Authors are thankful to Director, CSIR-CBRI, Roorkee for granting permission to publish this paper.

References

1. Cardinali, M., Reichenbach, P., Guzzetti, F., Ardizzone, F., Antonini, G., Galli, M., Cacciano, M., Castellani, M. andSalvati, P. 2002. A geomorphological approach to the estimation of landslide hazards and risks in Umbria, Central Italy. *Nat Hazards Earth Syst Sci*, 2: 57–72.

2. Cruden, D.M. andVarnes, D.J. 1996. Landslide types and processes. In: Turner AK, Schuster RL (eds.), Landslides: investigation and mitigation (Special Report 247). Washington, DC: National Research Council, Transportation and Research Board, pp. 36–75.

3. Gupta, R.P., Kanungo, D.P., Arora, M.K., Sarkar, S. 2008. Approaches for comparative evaluation of raster GIS-based landslide susceptibility zonation maps. *International Journal of Applied Earth Observation and Geoinformation*, 10: 330-341.

4. Kanungo, D.P. and Sharma, S. 2014. Rainfall thresholds for prediction of shallow landslides around Chamoli-Joshimath region, Garhwal Himalayas, India. *Landslides*, 11: 629-638.

5. Kanungo, D.P., Arora, M.K., Sarkar, S., Gupta, R.P. 2009. A fuzzy set based approach for integration of thematic maps for landslide susceptibility zonation. *Georisk*, 3: 30-43.

6. Kanungo, D.P., Arora, M.K., Sarkar, S., Gupta, R.P. 2006. A comparative study of conventional, ANN black box, fuzzy and combined neural and fuzzy weighting procedures for landslide susceptibility zonation in Darjeeling Himalayas. *Engineering Geology*, 85: 347-366.

7. Li, Z., Nadim, F., Huang, H., Uzielli, M. andLacasse, S. 2010. Quantitative vulnerability estimation for scenario based landslide hazards. *Landslides*, 7: 125-134.

8. Sarkar, S., Kanungo, D.P. and Sharma, S., 2015. Landslide hazard assessment in the upper Alaknanda valley of Indian Himalayas. *Geomatics, Natural Hazards and Risk*, 6: 308–325.

Chapter 18

Influence of Matric Suction on Pullout Resistance of Soil Nails

P. Rajeevkaran* and A. Kulathilaka

Department of Civil Engineering,
University of Moratuwa, Sri Lanka
****E-mail: rajeevkaran1@live.com***

ABSTRACT

The pullout resistance of soil nail designs is computed using the conventional formulae ignoring the influence of matric suctionand is conservative. In most soil-nailing designs, nails are located above the ground water table and significant matric suctions would prevail, although the infiltration would cause some reduction. These values could be high in residual soil formations in tropical countries. Attempts were made in this research to understand the influence of matric suction on the pullout resistance of soil nails. In this research a soil nail was installed in a compacted residual soil where matric suction was monitored and varied by controlled wetting. Pullout resistance of the nail was determined under a matric suction of 40 kPa. The test was done in a test box of dimensions 1.3m × 1.08m × 0.96 m. The nail was installed at downward angle of 10° to the horizontal; an overburden pressure of 40 kN/m² was given by jacking against the loading frame. The result obtained was compared with numerical model in Plaxis 2D and design formulae.

Keywords: *Soil nail, Pullout resistance, Matric suction, Tensiometer, Plaxis 2D.*

1. Introduction

Soil nailing is a widely used slope stabilization technique that uses passive elements *i.e.* soil nails (reinforcement bar within a grouted body). Soil nailing is used to stabilize many civil engineering structures such as abutments, retaining walls, unstable slopes, *etc.* The use of soil nailing has increased in recent years due to its technical and economic advantages. When the soil mass attempts to move down, the tensile force mobilized in the nail will increase the shear resistance by increasing the normal stress along the potential failure surface and reducing the shear stress that has to be mobilized by the soil to maintain equilibrium.

The two key engineering features that should be considered in the design of soil nails are: (i) the pullout capacity of the nails generated by interface friction developed during various stages of loading, and (ii) the tensile capacity of the nails. Pullout capacity is affected by matric suction, overburden pressure, dilatancy, method of installation, angle of internal friction of soil, and grouting pressure. Soil nails are generally positioned in a region where the soil is in an unsaturated state and the soil nails are influenced by the matric suction in the soil mass. The influence of matric suction on the pullout capacity has not received much attention in the design of soil nails (Su *et al.*, 2008; Zhang *et al.*, 2009). Matric suction is a major parameter that contributes to the uncertainties in the estimation of the pullout capacity of soil nails (Zhang *et al.*, 2009).

Matric suction pressure is the difference between pore air (Ua) and pore water (Uw). The matric suction reduces with the level of saturation as air pores become filled with water. This leads to a loss of shear strength of the soil. Soil nailing applications are mostly done above the ground water table, where the soil is in an unsaturated condition. In most designs the contribution of matric suction or the negative pore-water pressures in the vadose zone is neglected. Pullout capacity of clay gravel is reduced to 50 per cent when the water content was increased from optimum water content to full saturation (Schlosser *et al.*, 1983). The test results obtained by Su et. al. (2008) showed that the peak pullout strength of soil nails was strongly influenced by the degree of saturation of soil. There are several methods currently being used to estimate the pullout resistance of soil nails. Most of the formulae do not account for the presence of matric suction and the nature of shear strength parameter variation with saturation (matric suction). However, the method proposed by Gurpersaud (2010) has accounted for the matric suction and allowed for the dilation effect as well.

$$Q_{f(us)} = [(c_a + \beta\sigma'_z) + \{(u_a - u_w)(S^K)\tan(\delta + \Psi)\}]\pi dL$$

$$\beta = k_\theta \tan(\delta + \Psi),\ k_\theta/k_0 = 1 + (1 - k_0)/2k_0\ x(1/\cos 2\theta)$$

where,

$Q_{f(us)}$ = ultimate pull-out resistance (kN), β = Bjerrum-Burland coefficient, S = degree of saturation, c_a= apparent cohesion of the soil, K = fitting parameter, Ψ = dilation angle, θ = cutting slope, σ_v = effective vertical stress calculated at the mid-point of the nail in the resistance zone, δ = interface friction angle, k_θ = coefficient of lateral earth pressure, k_0 = coefficient of lateral earth pressure at rest, L = effective length of the soil nail, d = diameter of the nail.

The main objective of this study is to understand how the matric suction influences the development of pullout resistance of grouted soil nails.

2. Material and Methods

2.1 Soil and Material Properties

The characteristics of the soil used to make the compacted fill was determined by sieve analysis, hydrometer analysis, Atterberg limit test, specific gravity test and the standard Proctor compaction test. According to USCS (Unified Soil Classification System), the soil is silt with high plasticity (MH). **Table 18.1** summarizes the properties of soil.

Table 18.1: Properties of Soil

Type of Soil	*Silt of High Plasticity (MH)*
Optimum moisture content	23.10 per cent
Maximum dry density	1554 kg/m^3
Specific gravity	2.61
Liquid limit	55.80 per cent
Plastic limit	36.35 per cent
Plastic index	19.45 per cent

2.2 Detail of Equipments

To ensure uniform conditions, soil was compacted in layers in a box and two soil nails were installed in the fill. The dimension of the Perspex box is 1.3×1.08×0.96m. A clear distance of more than 200mm was allowed in all sides to avoid the influence of boundary effects following the suggestion by Zhou *et al.* (2011). The grouted nailof the study haddiameter 100 mm and length 914 mm and was fixed with an inclination of 10° downward to the horizontal. The test box layout is illustrated in **Figure 18.1**.

2.3 Compaction of Soil Mass

Water was added to the loose soil to bring it to optimum moisture content, mixed well and left for a night to bring it to moisture equilibrium. Then it was placed in layers of thickness 125 mm and the compaction was done manually with a modified hammer providing an energy level equivalent to standard Proctor. After compaction of each layer the relative compaction was evaluated by measuring the in situ density with a modified core cutter which has a diameter of 44 mm and height of 109 mm. Relative compaction of 90 - 100 per cent was achieved in all layers.

2.4 Drilling

The hole for the nailing was done by manual auguring with augers of diameter 75 to 100mm. First, it was drilled with 75 mm augur and the nail alignment was corrected. Then a 100 mm core augur was used and diameter correction was done. The length of the hole was 1.12m from the face of the box but only the lower 0.914m was grouted.

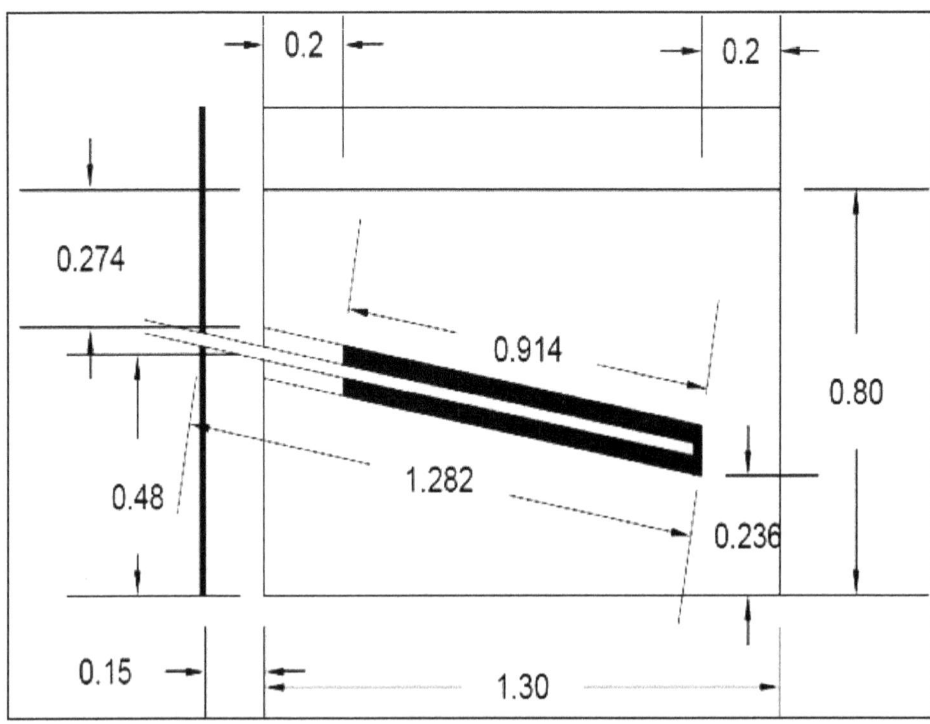

Figure 18.1: Dimensions of Test Box with Soil Nail (in meters).

2.5 Grouting

The grouting was done using a manual syringe made of PVC pipe with a diameter of 50 mm. The grout was sucked into the syringe first and then it was pushed through an additional pipe from the end to the front, progressively removing the additional pipe. Tokyo super ordinary portland cement and flow cable 50 in the mix proportion of 200:1 was used for grouting. Reinforcement bar of 25 mm QstRb steel was used.

2.6 Instrumentation

The force applied and the displacement of the nail during the pullout test was recorded by a hollow jack and the dial gauge fixed at the end of the nail. **Figure 18.2** and **18.3** shows the pullout test arrangement. The matric suction was measured at two locations along the nail continuously by tensiometers and all were connected to a computer through channel box and data logger.

2.7 Test Procedure

The test sequence that followed after compaction is:

- The soil nail was installed with manual drilling and grouting.
- The suction was measured in the compacted soil at two places and the soil mass was wetted until the matric suction reduced to a measurable level (less than 86 kPa).

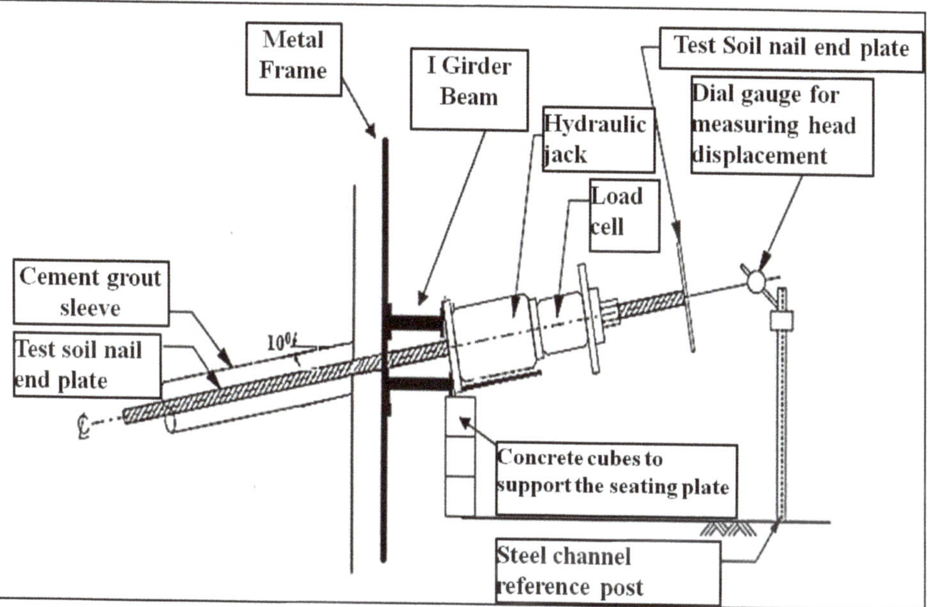

Figure 18.2: Pullout Test Arrangement.

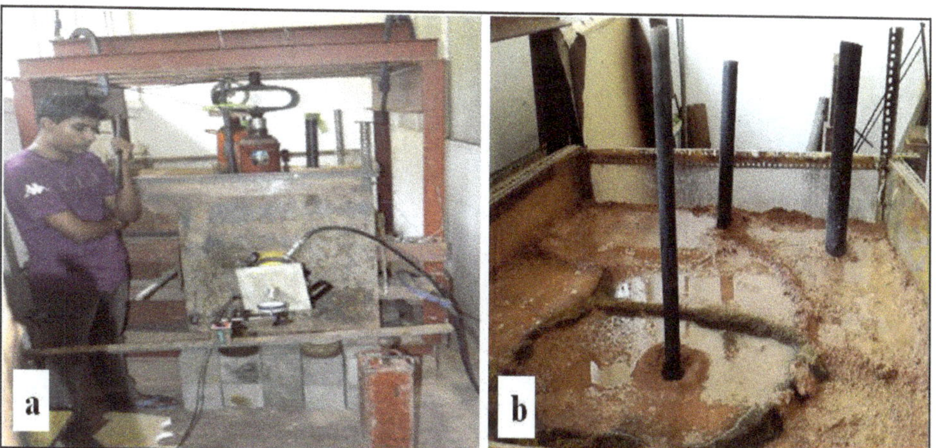

Figure 18.3: Pullout Testing (a) The Pullout Testing and (b) Saturation of Soil.

- The pullout test was conducted by the force controlled method in a controlled manner applying the pullout force in increments of 1 kN.
- For every 1 kN increments displacements were taken at 0, 1, 2, 5 minutes holding time. The pullout force was measured by a load cell with a digital output.
- The force was increased up to failure.

2.8 Numerical Analysis

The laboratory study was numerically simulated in Plaxis 2D software using the plane strain model. Standard fixity was used for the boundary condition. The grouted soil nail was modeled as plate with reference to Babu and Singh (2009). An overburden pressure of 40 kPa was applied on the top of the soil mass. The Mohr-Coulomb failure criterion was used for analysis and a prescribed displacement of 13.1 mm was given to the nail.This was the actual displacement observed in the test. Therefore, the actual process of model construction and testing was closely simulated. **Figure 18.4** shows the model drawn in Plaxis 2D and **Table 18.2** shows (a) soil and (b) grouted nail properties used in the analysis.

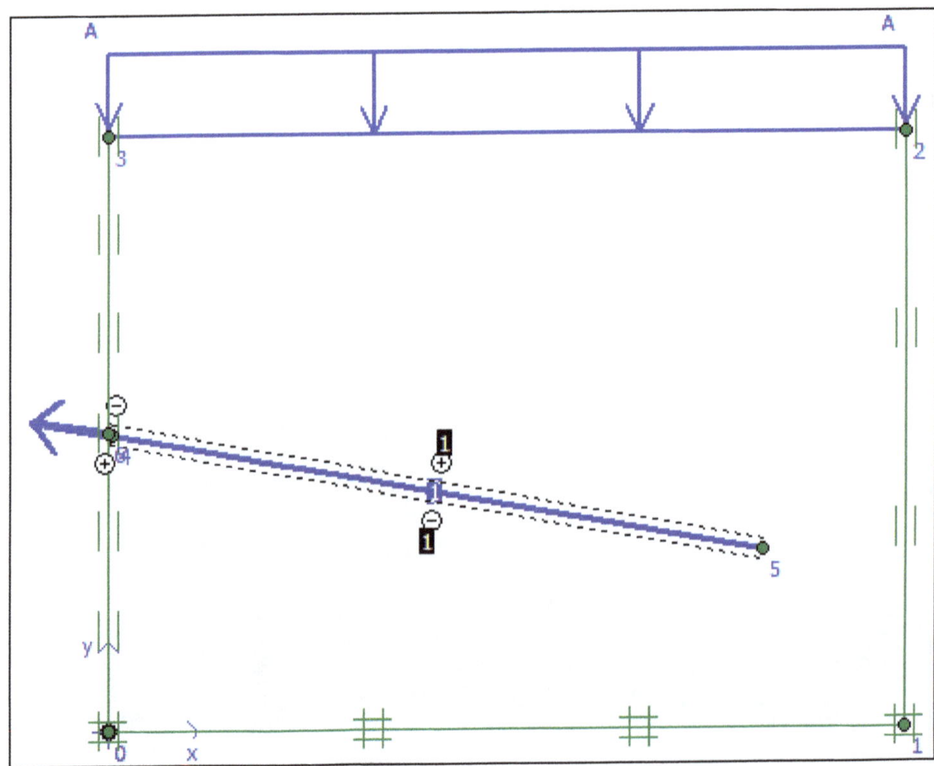

Figure 18.4: Model Drawn in Plaxis 2D.

3. Results and Discussions

The average matric suction was obtained by linear consideration of the matric suction profile. The nail was pulled out at a matric suction of 39.88 kPa. The pullout force was increased up to a peak of 11.72 kN and the displacement kept at 13.10 mm in the actual test. The pullout load obtained from the numerical analysis was approximately 11.41kN. The results of experiment and numerical stimulation are quite close. The estimated pullout resistance for the test based on the Gurpersaud (2010) formula, using the parameters of the compacted fill (S = [illegible] per

Table 18.2: Showed (a) Soil and (b) Grouted Nail Properties Used in the Analysis

(a) Soil Properties	
Residual soil	
Cohesion (kPa)	46
Internal friction angle (deg)	32.62
Unit weight (kN/m³)	17
Elastic modulus (MPa)	16
Poisson's ratio of soil	0.3
Interface strength	0.75
(b) Grouted Nail Properties	
Grouted nails	
Material model	Elastic
Elastic modulus of reinforcement (GPa)	200
Elastic modulus of grout (GPa)	22
Diameter of reinforcement bar (mm)	25
Drill hole diameter (mm)	100
Length of nail (mm)	914
Declination wrt Horizontal (deg)	10
Spacing (mm)	350

cent and assumed k = 1) are 13.45 kN. It will be 9.05 kN undersaturated condition (conventional formula). The comparison pullout results obtained through different methods are tabulated in **Table 18.3. Figure 18.5a** shows the result of pullout force versus deflectionat 5 minute sholding time and **Figure 18.5b** shows the total displacement vectors after simulation of pullout.

Table 18.3: Pullout Results Obtained through different Methods

	Experiment	*Plaxis Modelling*	*Conventional Formula*	*Gurpersaud Formula*
Pullout force (kN)	11.72	11.41	9.05	13.45

4. Conclusion and Recommendation

A soil nail was installed in a compacted residual soil and a pullout test was conducted under 40 kPa matric suction. The pullout capacity of soil nail placed in unsaturated residual soil is 1.3 times higher than soil nail in saturated soil. This result shows that there is a strong relationship between matric suction and the pullout capacity of the soil nail installed in residual soil. Additional testing is necessary to verify the influence of matric suction on the pullout capacity of soil nails.

Acknowledgements: Assistance extended by the Laboratory Staff members of Engineering Laboratory Service ELS (Pvt.) Ltd during the installation and pullout

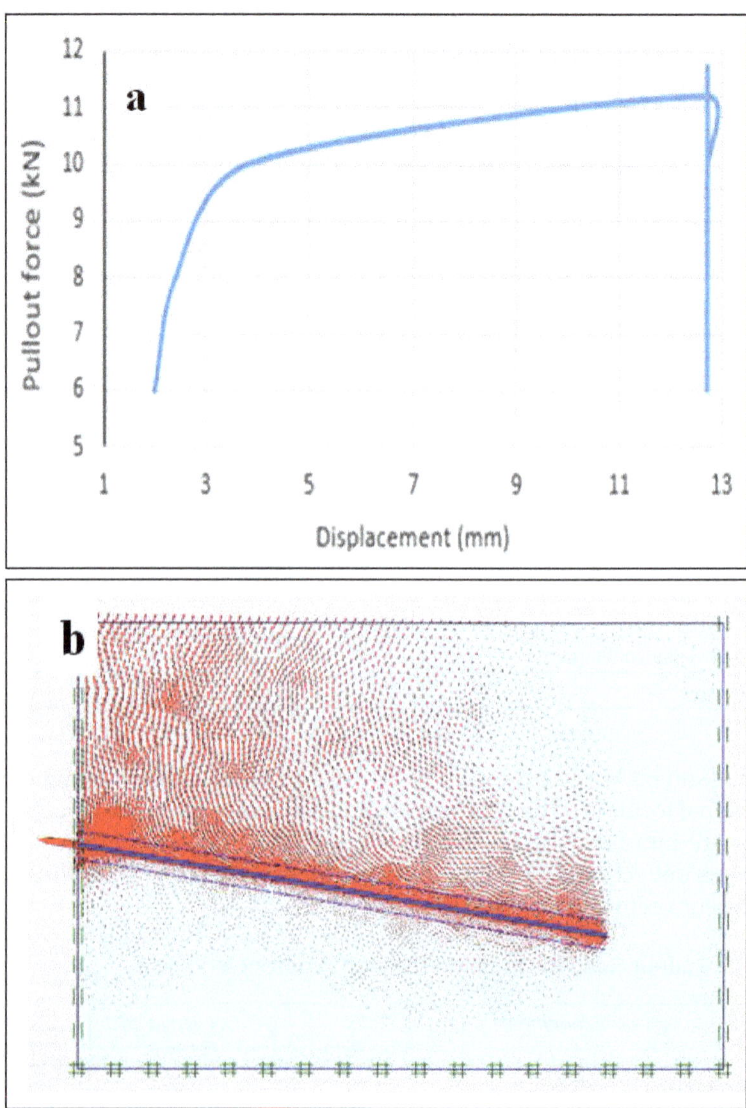

Figure 18.5: (a) Pullout Force versus Deflection Curve, and (b) Total Displacement after Simulation.

testing of the nails is gratefully acknowledged. Support received from laboratory staff at the Soil Mechanics laboratory is also appreciated.

References

1. Babu G.L.S. and Singh V.P. 2009. Plaxis practice - Simulation of soil nail structures using PLAXIS 2D. *Plaxis Bulletin*, Spring issue, pp.16-21.

2. Gurpersaud N. 2010. The influence of matric suction on the pull-out capacity of grouted soil nails. M.Sc Thesis, University of Carleton, pp.8-30

3. Schlosser F., Jacobsen H.M. and Juran I. 1983. Soil reinforcement. General Report Specialty Session 5. In: *Proceedings of the 8th International Conference in Soil Mechanics and Foundations Engineering*. Helsinki, pp.1159-1180.
4. Su L.J., Chan T.C.F., Shiu Y.K., and Cheung T. and Yin J.H. 2008. Influence of degree of saturation on soil nail pullout resistance in compacted completely decomposed granite fill. *Canadian Geotechnical Journal* 44: 1314-1328
5. Zhang L.L., Zhang L.M., and Tang W.H. 2009. Uncertainties of field pullout resistance of soil nails. *Journal of Geotechnical and Geoenvironmental Engineering* 135: 966-973
6. Zhou W.H., Yin J.H. and Hong C.Y. 2011. Finite element modeling of pullout testing on a soil nail in a pullout box under different overburden and grouting pressures. *Canadian Geotechnical Journal* 48: 557-567.

Section V

Earthquakes

Chapter 19

The SSA-GEM Seismic Hazard Model for the East African Rift

V. Poggi[1], R. Durrheim[2], G.M. Tuluka[3], G. Weatherill1[1], R. Gee[1,6], M. Pagani[1], A. Nyblade[4] and D. Delvaux[5]*

[1]Global Earthquake Model Foundation (GEM), Via Ferrata 1, 27100 Pavia, Italy
[2]University of the Witwatersrand, Braamfontein 2000, Johannesburg, South Africa
[3]Goma Volcanic Observatory, mt. Goma, DR Congo
[4]Penn State University, University Park, Pennsylvania USA
[5]Royal Museum for Central Africa, Tervuren, Belgium
[6]Istituto Nazionale di Oceanografia e di GeofisicaSperimentale (OGS), Borgo Grotta Gigante, Sgonico (TS), Italy
**E-mail: raymond.durrheim@wits.ac.za*

ABSTRACT

The East African Rift System (EARS) is the major active tectonic feature of the Sub-Saharan Africa (SSA) region. Although the seismicity level of this divergent plate boundary can be described as moderate, several damaging earthquakes have occurred in historical times. The seismic risk is exacerbated by the high vulnerability of the local buildings and structures. We present an updated seismic hazard model for the region.

Keywords: *Seismic hazard, East African Rift, Global earthquake model, Open quake.*

1. Introduction

Sub-Saharan Africa is largely a stable intra-plate region characterized by a relatively low level of seismic activity, with earthquakes randomly distributed in space and time (**Figure 19.1**). The only exceptions are the *East African Rift System*

(EARS) and the Cameroon Volcanic Line, where earthquakes are associated with active fault zones and volcanic activity, respectively. Damaging earthquakes with M>6 occur almost annually in the EARS. Five M>7 earthquakes have occurred in eastern Africa since nineties, the largest known event being the 13th Dec. 1910 Ms 7.4 Rukwa (Tanzania) that badly cracked all European-style houses in towns on the eastern shore of Lake Tanganyika (Ambraseys 1991, Midzi and Manzunzu 2014). While these events caused relatively small losses, the population has increased enormously over the last century, and traditional buildings, which have a large inherent resistance to earthquake shaking, have been replaced byvulnerable European-style unreinforced masonry. The occurrence of similar events close to a town would likely now cause serious human and economic losses. In the 21st century there have been several events that have caused loss of life (Durrheim 2016). Hence Africans cannot be complacent: a damaging earthquake could occur anywhere, although the frequency is greatest in tectonically active regions such as the East African Rift.

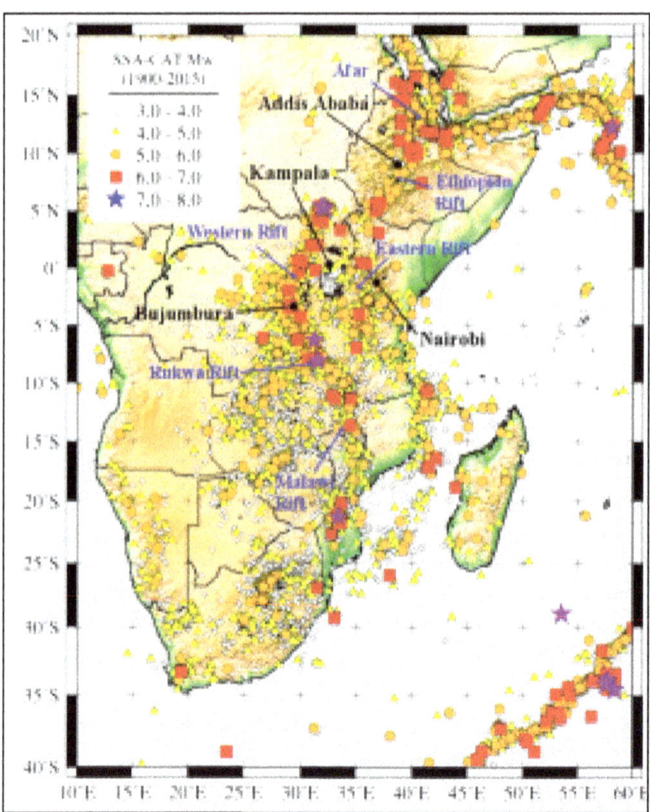

Figure 19.1: $M_w \geq 3$ Earthquakes in the Homogenized SSA-GEM Catalogue. Blue labels - major rift systems; Black labels-African capitals analyzed in this study.

2. Probabilistic Seismic Hazard Assessment (PSHA)

The most recent probabilistic seismic hazard assessment (PSHA) covering Africa was published by the Global Seismic Hazard Assessment Program (GSHAP) (Grünthal *et al.*, 1999, Midzi *et al.*, 1999), based on a catalogue that extended to 1996. In this USAID-funded project we extended the GSHAP catalogue and improved the PSHA map using Global Earthquake Model (GEM) tools, products and data from temporary deployments of seismic networks in an effort to reduce seismic risk associated with the East African Rift (Poggi *et al.*, 2017, **Figure 19.2**).

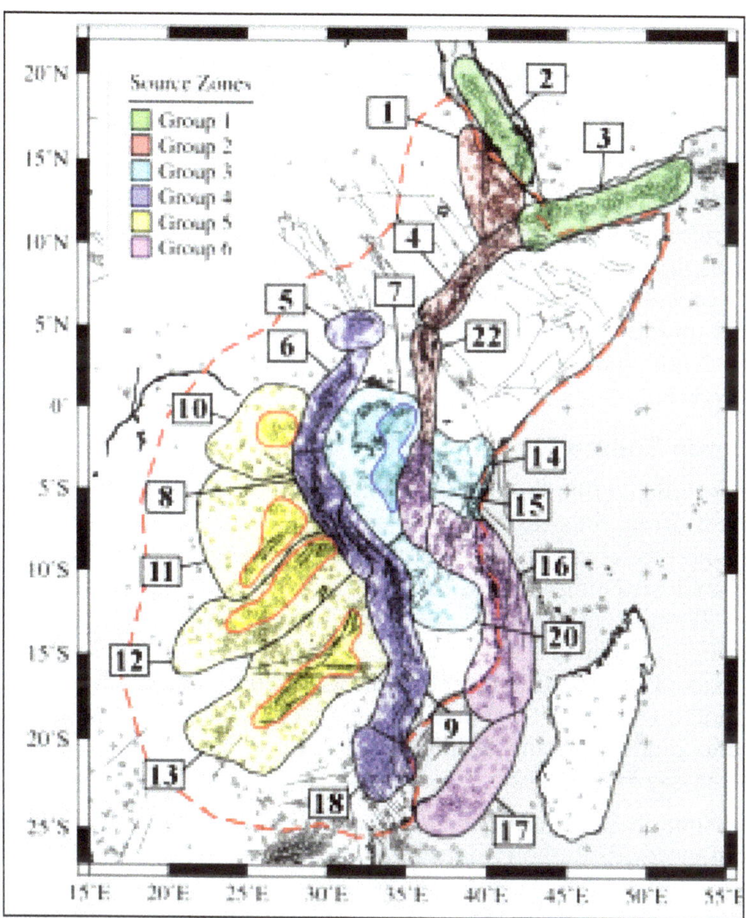

Figure 19.2: SSA-GEM Source Zonation Model. The red dashed line marks the PSHA calculation area. The faults are from MacGregor (2015).

2.1 GEM-SSA Catalogue

The manually-reviewed bulletin from the International Seismological Centre (ISC, 2013) was the primary source of information for the earthquake catalogue. In our selected geographic area (-40° to 20°N, 10° to 60°E; **Figure 19.1**) it spans the period 1904-2013, and includes a total of 26,322 events. The ISC-GEM global instrumental earthquake catalogue (Storchak *et al.*, 2013, 2015) is a refined version of the ISC bulletin, which improves the accuracy of magnitude and location solutions for large global events (M_w>5.5) in the period 1900-2012. Ver.3 of the ISC-GEM global instrumental catalogue was used in this study, which has285 events within the study region.

The GEM Historical Earthquake Catalogue is a global collection of reviewed historical records consisting of 825 events (M>7) covering the period 1000-1903. Only eight earthquakes in the catalogue fall within the study region. We supplemented the earthquake record with local catalogues compiled from three campaigns: (i) the Tanzanian Broadband Seismic Experiment (2,218 events; 1994-1995; Langston *et al.*, 1998); (ii) the Ethiopian Plateau Catalogue (253 events, 2001-2002; Brazier *et al.*, 2008); and (iii) the Africa Array Eastern Africa Seismic experiment (1,023 events; 2009-2011; Mulibo and Nyblade 2016). The result is the Sub-Saharan Africa – Global Earthquake Model (SSA-GEM) catalogue.

The GEM catalogue toolkit (Weatherill *et al.*, 2016) was used to homogenize magnitudes. Events recorded in more than one catalogue falling within 0.5° and 120s were considered duplicates. We used the original magnitude-scaling relation of Gardner and Knopoff (1974) to identify and remove fore- and aftershocks. The declustered SSA-GEM catalogue consists of 7,259 $3 \leq M_w \leq 7.5$ events out of the original 29,803.

2.2 Seismic Source Model

The seismic source model for Sub-Saharan Africa is based on distributed seismicity sources, consisting of areal source zones (ASZ) representing uniform temporal and spatial earthquake occurrence. Seismicity constraints were obtained from the analysis (completeness, occurrence rates) of the SSA-GEM earthquake catalogue. Tectonic information was derived mostly from scientific literature and by integration of available datasets. We defined 20 source zones stretching from the Red Sea and Gulf of Aden, through the Main Ethiopian Rift, branching around the Tanzanian Craton with splays into the Democratic Republic of the Congo, Zambia and Zimbabwe, and terminating in Mozambique, the Indian Ocean and Madagascar (see **Figure 19.2**).

The source depth distribution model was derived from the SSA-GEM catalogue. The overall strike distribution was calibrated by performing statistical analysis on the outcropping fault structures database (MacGregor 2015). Catalogue completeness was evaluated for different temporal periods and magnitude ranges using the step algorithm as implemented in the GEM's Hazard Modeller Toolkit (HMTK) (Weatherill 2014). Seismicity in each area source was assumed to follow a double-truncated Gutenberg-Richter magnitude-frequency distribution.

2.3 Logic Tree Implementation

Ground Motion Prediction Equations (GMPEs) are important in assessing seismic hazard. However, Sub-Saharan Africa suffers from a severe lack of data. The use of temporary networks did not contribute significantly, as no large magnitude events were recorded and recordings in the near- to intermediate distance range (<50 km) were lacking. Consequently we selected GMPEs based on factors such as the tectonic setting, the type and quality of data used for calibration, and the suitability of the functional form (Cotton *et al.*, 2006). The current logic-tree model was restricted to the use of four GMPEs, two for active shallow crust (Chiou and Youngs 2014, Akkar *et al.*, 2014) and two for stable continental conditions (Atkinson and Boore 2006, Pezeshk *et al.*, 2011). Logic-tree weights were based on the assessment of local seismotectonic conditions by a pool of experts from the region.

2.4 Hazard Computation

Probabilistic Seismic Hazard Assessment (PSHA) computations were performed using the Open Quake-engine (Ver. 2.0; Pagani *et al.*, 2014). The investigation area consists of a mesh of 79,109 sites spaced at approximately 10 km. For each site of the mesh, free rock conditions are assumed, with a fixed 30-metre averaged shear-wave velocity (Vs 30) reference of 600 m/s. In order to predict the earthquake response of buildings and other structures, the probability was computed of exceeding a range of Peak Ground Accelerations (PGA) and Spectral Accelerations (SA) for the response spectral periods of 0.05, 0.1, 0.2, 0.5, 1 and 2 seconds. The target ground motion intensity for calculation was 5 per cent damped response spectral acceleration (in g), estimated for probabilities of exceedance (PoE) of 10 per cent and 2 per cent

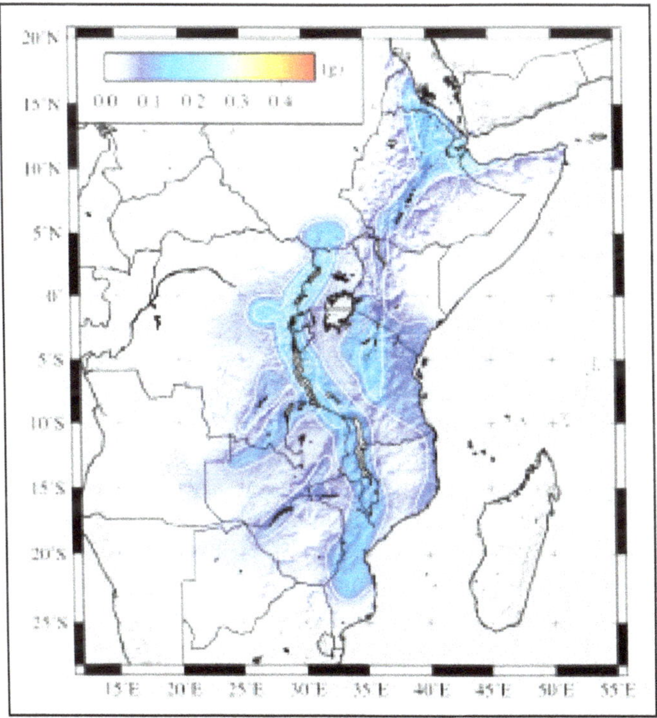

Figure 19.3: Map of Peak Ground Acceleration (PGA, in g) for a 10 per cent Probability of Exceedance in 50 Years.

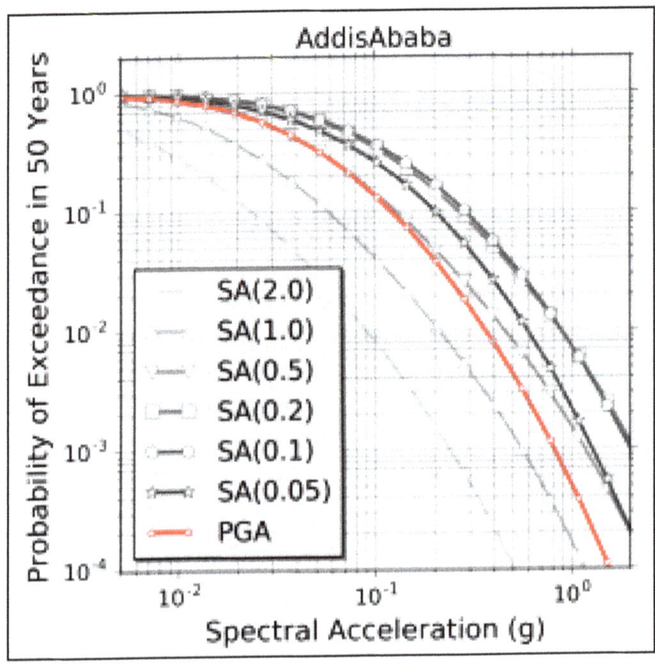

Figure 19.4: PGA and Mean Hazard Curves at Addis Ababa Computed for Spectral Response Periods Ranging from 0.05-2 s.

within an investigation time of 50 years, the spectral acceleration was computed at each grid point (see **Figure 19.3**) and for several capital cities (see **Figure 19.4** shows the mean hazard curves for Addis Ababa).

3. Conclusions

The SSA-PSHA model is generally consistent with the previous regional model from the GSHAP project (Midzi *et al.*, 1999), with some noticeable differences. The biggest difference (SSA model 0.08 g smaller) is found in the south Sudan cluster (Juba region), which is likely due to the different modelling strategy of the area sources. A similar situation is found towards the south, where the SSA model predicts a slightly lower acceleration for the Lake Tanganyika region (0.2 g) compared to the region of the northern lakes (Kivu, Edward and Albert). Conversely, the SSA model of the southern tail of the western branch (*e.g.* in Malawi) shows a considerably higher acceleration (0.15 g) than GSHAP (0.08 g). These differences are explained by the expanded catalogue and different calibration of seismicity parameters.The major shortcoming of the SSA model is the lack of strong-motion recordings to help select and validate existing ground motion prediction models. Improvements in the design of the logic tree structureare also possible. Up to now, the only considered epistemic variability of the source model is the uncertainty in M_{max}, while any error in b-value and occurrence rate is neglected.

Acknowledgements

GMT and RJD thank the Global Earthquake Model Foundation for training and support in using the Hazard Modeller's Toolkit and the OpenQuake Engine and for their kind hospitality in Pavia, Italy. RJD acknowledges the support of the South African Research Chairs Initiative.

References

1. Akkar S., Sandikkaya M.A. and Bommer J.J. 2014. Empirical ground-motion models for point- and extended-source crustal earthquake scenarios in Europe and the Middle East. *Bulletin of Earthquake Engineering* 12: 359-387
2. Ambraseys N.N. 1991.The Rukwa earthquake of 13 Dec. 1910 in E. Africa. *Terra Nova* 3: 203-208
3. Atkinson G. and Boore D. 2006. Earthquake ground-motion prediction equations for eastern North America. *Bulletin of the Seismological Society of America* 96: 2181-2205
4. Brazier R.A., Miao Q., Nyblade A.A., Ayele A., and Langston C.A. 2008. Local magnitude scale for the Ethiopian Plateau. *Bulletin of the Seismological Society of America* 98: 2341-2348
5. Chiou B.S.-J. and Youngs R.R. 2014.Update of the Chiou and Youngs NGA model for the average horizontal component of peak ground motion and response spectra. *Earthquake Spectra* 30: 1117-1153
6. Cotton F., Scherbaum F., Bommer J.J. and Bungum H. 2006. Criteria for selecting and adjusting ground motion models for specific target regions: Application to central Europe and rock sites. *Journal of Seismology* 10: 137-156
7. Durrheim R.J. 2016. African Earthquakes. In Natural Hazards and Disasters in sub-Saharan Africa. G. Mulugeta and T. Simelane (Eds), Africa Institute of South Africa, Pretoria.
8. Gardner J. K. and Knopoff L. 1974. Is the sequence of earthquakes in Southern California, with aftershocks removed, Poissonian? *Bulletin of the Seismological Society of America* 64: 1363-1367
9. Grünthal G., Bosse C., Sellami S., Mayer-Rosa D., and Giardini D. 1999. Compilation of the GSHAP regional seismic hazard for Europe, Africa and the Middle East. *Annals of Geophysics* 42: 1215-1223
10. ISC 2013. International Seismological Centre. On-line Bulletin, http://www.isc.ac.uk, IntSeismol Cent, Thatcham, United Kingdom.
11. Langston C.A., Brazier R., Nyblade A.A. and Owens T.J. 1998. Local magnitude scale and seismicity rate for Tanzania, East Africa. *Bulletin of the Seismological Society of America* 88: 712-721.
12. MacGregor D. 2015. History of the development of the East African Rift System: a series of interpreted maps through time. *Journal of African Earth Sciences* 101: 232-252.

13. Midzi V., Hlatywayo D.J., Chapola L.S., Kebede F., Atakan K., Lombe D.K., Turyomurugyendo G., and Tugume F.A. 1999. Seismic hazard assessment in eastern and Southern Africa. *Annals of Geophysics* 42: 1067-1083

14. Midzi V. and Manzunzu B. 2014. Large recorded earthquakes in sub-Saharan Africa: In Extreme Natural Hazards, Disaster Risks and Societal Implications, eds. A. Ismal-Zadeh, J. Urrutia-Fucagauchi, A. Kijko and I Zaliapin. Cambridge: Cambridge University Press.

15. Mulibo G.D. and Nyblade A.A. 2016. The seismotectonics of southeastern Tanzania: implications for the propagation of the Eastern Branch of the East African Rift. *Tectonophysics* 974: 20-30.

16. Pagani M., Monelli D., Weatherill G.A. and Garcia J. 2014. The OpenQuake-engine Book: Hazard. Global Earthquake Model (GEM) Technical Report 2014-08, doi: 10.13117/-GEM.OPENQUAKE.TR2014.08, pp.67

17. Pezeshk S., Zandieh A. and Tavakoli B. 2011. Hybrid empirical ground-motion prediction equations for eastern North America using NGA models and updated seismological parameters. *Bulletin of the Seismological Society of America* 101: 1859-1870

18. Poggi V., Durrheim R., Mavonga Tuluka G., Weatherill G., Gee R., Pagani M., Nyblade A. and Delvaux D. 2017. Assessing seismic hazard of the East African Rift: a pilot study from GEM and Africa Array. *Bulletin of Earthquake Engineering* 15: 4499-4529

19. Storchak D.A., Di Giacomo D., Bondár I., Engdahl E.R., Harris J., Lee W.H.K., Villaseñor A. and Bormann P. 2013. Public release of the ISC-GEM Global Instrumental Earthquake Catalogue (1900-2009). *Seismological Research Letters* 84: 810-815.

20. Storchak D.A., Di Giacomo D., Engdahl E.R., Harris J., Bondár I., Lee W.H.K., Bormann P. and Villaseñor A. 2015. The ISC-GEM Global Instrumental Earthquake Catalogue (1900-2009). *Physics of the Earth and Planetary Interiors* 239: 48-63

21. Weatherill G.A. 2014. OpenQuake Hazard Modeller's Toolkit - User Guide. Global Earthquake Model Technical Report.doi:10.13117/gem.openquake.man.hmtk.01.

22. Weatherill G.A., Pagani M. and Garcia J. 2016. Exploring earthquake databases for the creation of magnitude-homogeneous catalogues: tools for application on a regional and global scale. *Geophysical Journal International* 206: 1652-1676

Section VI

Integrated Disaster Risk Reduction

Chapter 20

The Impacts of Extreme Natural Events: S&T Awareness, Development and Education in Myanmar

M.S. Aung

Department of Research and Innovation,
No.6, Kabaraye Pagoda Road, Yangon, Myanmar
E-mail: dr.myatsoeaung@gmail.com

ABSTRACT

Even though science and technology can mitigate human disasters caused by extreme natural events, developing country like Myanmar cannot utilize it effectively to reduce such events. Myanmar is one of the largest country in South East Asia with 57 million people. Myanmar has made significant progress in its disaster management policy, plans, and procedures since 2008, when Cyclone 'Nargis' impacted the country leaving disastrous impact. Over the last decades, Myanmar has faced two major earthquakes, three severe cyclones, floods, and other smaller scale natural calamities. So, health and education program were implemented for health care system and capacity building and training for such natural disasters. When S&T use in disaster risk reduction, it has barrier including lack of political and public awareness, inadequate institutional mechanisms and technical capacities and of course absence of sustainable funding. It may difficult to apply S&T in situation of chronic or extensive risk. However, Myanmar Action Plan on Disaster Risk Reduction (MAPDRR) is developed and endorsed in 2012 by Myanmar Government. MAPDRR is in line with the Hyogo Framework for Action (HFA) and the ASEAN Agreement on Disaster Management and Emergency Response (AADMER). Myanmar Disaster preparedness management working committee have been organized for National Disaster Management. Myanmar introduced Disaster Management Law for DRR commitments. A comprehensive training is given to young people who are selected from hazardous areas and helping communities to prepare themselves for any such future events and ways to decrease. DRI and MES provide advice on scientific and technical issues related to the reduction of disaster risks, implementation and also assist the coordination of scientific and technical activities. Myanmar is planning to establish monitoring stations. But Instruments, human resources and capacity building are needed to upgrade. This paper will gives us an overall understanding of disaster risk reduction in Myanmar concerning with S&T awareness, development and education.

Keywords: *MAPDRR, HFA, AADMER, DRR, Disaster management law.*

Abbreviations

STI: Science, Technology and Innovation
MoE: Ministry of Education
NDPCC: National Disaster Preparedness Central Committee
HFA: Hyogo Framework for Action
ACDM: ASEAN Committee on Disaster Management
AADMER: ASEAN Agreement on Disaster Management and Emergency Response
ADPC: ASEAN Disaster Preparedness Center
DRR: Disaster Risk Reduction
MAPDRR: Myanmar Action Plan on Disaster Risk Reduction
ToT: Training of Trainers
DMTC: Disaster Management Training Centre
EOC: Emergency Operation Centre
DMH: Department of Meteorology and Hydrology
MNBC: Myanmar National Building Code

1. Introduction

Myanmar is exposed to multiple natural hazards in which coastal regions are exposed to cyclones, storm surges and tsunamis while most parts are at risk from earthquakes and fires. The rainfall-induced flooding is across the country while some parts are exposed to landslide and drought risks. Myanmar is striving for a peaceful, modern and developed nation but natural disasters destroy the developmental gains and hinder the developmental interventions. The preparedness and mitigation should be an integral part of the development plans. It is also important to make prior arrangement for relief, rehabilitation and reconstruction activities, in case a natural disaster strikes. Science, technology and innovation (STI) have emerged as the major drivers of national development globally. The importance of science and technology in better understanding the processes before, during and after disasters is becoming increasingly. An increasing number of the world's population would be impacted by a climate-based disaster. Scientific data and information and the tangible application of technology is critical to under pinning well-informed policies and decisions across the public, private and voluntary sectors. Much scientific evidence exists but better links to decision-making in policy and planning are needed to continuously enhance our ability to forecast, reduce and respond to disaster risks. Science and technology can assist in identifying a problem, developing understanding from research, informing policy and practice and making a difference that can be objectively demonstrated when evaluated. Ministry of Education (MoE) arranged that Students of all ages can study and participate in school safety measures, and also work with teachers and other adults in the community towards minimizing risk before, during and after disaster events. Education system is significantly impacted in disaster risk reduction and emergency preparedness as an integral part of education. Education is also one of the development activities in which children have to understand past disasters and their impact. The objective of this paper is to understand how Myanmar implementing in disaster risk reduction concerning with S&T awareness, development and education.

2. Information Analysis

2.1 Disaster Risk Profile of Myanmar

Myanmar's climate is largely tropical with three seasons- (1) the monsoon/rainy season (June to October), (2) cool season (November to February) and (3) hot season (March to May). Myanmar is exposed to a plethora of natural calamities, including earthquakes, fires, droughts, floods, landslides, cyclones and tsunamis. A total of 27 natural disasters have been recorded between 1980 and 2010, causing the death of approximately 140,000 people, and affecting the lives and livelihoods of 3.9 million people. During 2002-2012, three cyclones affected over 2.6 million people, floods affected over 500,000 people; two major earthquakes affected over 20,000 people. Almost the entire country is affected by natural calamities, with varying intensity depending on the hazard. Drought is the most persistent calamities throughout the country; cyclone impact three distinct regions of the country; earthquakes and floods significantly affect similar areas when weighted by mortality, however, floods generally causes more economic damage. Landslides also present a significant risk for regions on the western border.

2.2 Institutional Arrangement for Disaster Management in Myanmar

National Disaster Preparedness Central Committee (NDPCC) is the apex body for Disaster Management in Myanmar. National Disaster Preparedness Management Working Committee and Sub-Committee are constituted to supervise the implementation of Disaster Management activities under NDPCC. Division/State level committees have been constituted for effective implementation of disaster management activities (DRR Task Force of Government *et al.* (2009).

2.3 Global and Regional Disaster Risk Reduction Commitments

Myanmar is one of the 168 countries that endorsed the '***Hyogo Framework for Action***' (HFA) in 2005 aiming at "Substantial Reduction of Disaster Losses in Lives and in the Social, Economic and Environmental Assets of Communities and Countries". Myanmar is an active member of the ASEAN Committee on Disaster Management (ACDM) established in early 2003. Myanmar is also a signatory of the ASEAN Agreement on Disaster Management and Emergency Response (AADMER), which came into force in 2009. The AADMER is a proactive regional framework for cooperation, coordination, technical assistance, and resource mobilization in all aspects of disaster management, and the first legally binding HFA related instrument. Myanmar is a member of ASEAN Disaster Preparedness Center (ADPC) Regional Consultative Committee on Disaster Management since 2000 (DRR Task Force of Government *et al.* (2009).

2.4 Disaster Risk Reduction in Myanmar

Since 2011, Myanmar is pursuing a four wave's reform process; the political, economic and administrative reform and development of private sector, aiming to achieve political stability and economic development. The development partners and the international community welcome these changes and join in hands with the government. These will inevitably result to increase investments in infrastructure and rapid urbanization that encourages rural to urban migration. Meanwhile, if improvements in the development sectors do not integrate disaster risk reduction, they could exacerbate existing disaster risk and create new forms of disaster risk Building disaster resilience in Myanmar becomes more important than ever, to safe lives of Myanmar people, to protect investment and to ensure the sustainability of development gains (Centre for Excellence in Disaster Management and Humanitarian Assistance, January 2017).

2.4.1 Myanmar Action Plan on Disaster Risk Reduction (MAPDRR)

The Myanmar Action Plan on Disaster Risk Reduction (MAPDRR), that provides a framework for multi-stakeholder engagement on DRR in the country, was prepared with substantial consultation with various stakeholders. MAPDRR's goal is to make Myanmar safer and more resilient against natural hazards, thus protecting lives, livelihood and development gains. MAPDRR identifies 65 projects that need to be implemented to meet the Government's commitments to HFA and the ASEAN Agreement on Disaster Management and Emergency Response (AADMER).

2.4.2 Myanmar Disaster Management Law

The Government of Myanmar continued to demonstrate its commitment to DRR by introducing Myanmar Disaster Management Law that was enacted in July 2013. The Law includes the provisions for formation of disaster management bodies and their duties and responsibilities for all phases of disaster, establishment of disaster management fund at national and regional level. The law also provides the guidance to carry out the measures of disaster risk reduction along with the development plans in the country. The rules for implementation of Myanmar Disaster Management Law has been also drafted by the Ministry of Social Welfare, Relief and Resettlement in consultation with disaster risk reduction experts and Myanmar Disaster Risk Reduction Working Group composed of 56 international and national organizations, led by UNDP.

2.4.3 Youth Volunteer Network

Government has embarked on DRR Youth Volunteer Programme providing DRR related Training of Trainers (ToT) for young representatives from Ayeyarwaddy Region. Those young volunteer will be serving as a leader of community disaster management committee then to become a change agent who can promote the necessary change in behaviors. It plans to scale up the initiatives.

2.4.4 The Disaster Management Training Centre (DMTC)

Government approved the establishment of 'Disaster Management Training Centre (DMTC)' in order to build the capacity of people implementing disaster management activities. The DMTC has supported capacity development of officials from Government Departments and Social Organizations, since 1977. DMTC will be located in Hinthada Township in Ayeyarwaddy Region. The Ministry of Social Welfare, Relief and Resettlement is now undertaking DMTC to collaborate with international and local partners to mobilize the technical and financial resources for human resource and institutional capacity development, infrastructure development, development of curriculum and procurement of teaching aids and networking and partnership with International and Regional Training Institutes and Centers', Centre for Excellence in Disaster Management and Humanitarian Assistance (January 2017)

2.4.5 Emergency Operation Centre

With the aim to provide the supports for emergency management, response and logistic through information sharing on network and quick decision making, the Ministry of Social Welfare, Relief and Resettlement plans to set up Emergency Operation Centre (EOC).

2.4.6 Myanmar Disaster Loss and Damage Database

The Myanmar disaster loss and damage database has now been initiated by the Relief and Resettlement Department with the objective to develop national capacities for monitoring and analyzing risks and vulnerabilities to support disaster risk reduction, mitigation, preparedness, response and recovery. The National framework for the database has been finalized and the pilot data collection is now under-way. The database could be linked with the Regional and global networks in the near future.

3. Disaster Risk Reduction in Myanmar

Myanmar is affected by many natural hazards, destructive earthquake, cyclones, flooding, landslides and periodic droughts. Over the last decades, Myanmar has been impacted by two major earthquakes, three severe cyclones, floods and other small scale hazards. In 2015-16, the El-Niño phenomenon significantly impacted Myanmar so that it suffered extreme temperatures, unusual rainfall patterns, dry soil, high risk of fires and acute water shortages. Heavy rain started across the country in mid 2016, causing flooding and landslides in Rakine, Sagaing and Kachin. Myanmar has been faced to a wide range of disasters caused by various natural and human-made hazards. The high level of disaster risk is further compounded by climate change and variability, environmental degradation and haphazard development. Since the Cyclone Nargis catastrophe, Myanmar has accelerated programs meant to reduce and manage disaster risk. MAPDRR is aligned with the HFA and ASEAN Agreement on disaster management and emergency response, prioritizes seven components. In an environment of broad-based consultation and participation, the Risk Assessment Roadmap was developed. The risk assessment roadmap proposes several implementation mechanisms of disaster risk reduction in which analyzing potential hazards and evaluating existing conditions of vulnerability. These mechanisms are aligned with Myanmar Action Plan on Disaster Risk Reduction. The roadmap will implement within five years, by which the capacities and knowledge for risk informed decision making and development plans and thus will minimize the impacts of future disasters.

Moreover, Department of Meteorology and Hydrology (DMH) generates weather forecast and early warnings for cyclone, storm surge and flood. Color-coded cyclone warning message depend on disaster severity was started to use in 2009, as an attempt to make early warning message to be user friendly. Once the early warning on disaster is issued by DMH, it is informed to the public through media and TV. In order to improve the quality and accuracy of the weather forecast and early warning, DMH still needs to upgrade the capacity of equipment and tools for weather forecast. The DMH organizes Monsoon Forum as a mechanism for fostering a closer dialogue between forecast producers and users to enhance the uptake of weather and climate forecasts for disaster mitigation. The broader goal of the Forum is to build the national capacity to mitigate disaster risks by linking national hydro-meteorological agencies to sectors that are vulnerable to climate risks, notably agriculture, water resources, health, and disaster management project in collaboration with DMH. The use mobile phone SMS for early warning is also being explored. Greater priority of DRR should be put on sharing and disseminating scientific information, including technologicaladvances and translating them into practical methods that can readily be integrated into policies, regulations and implementation plans concerning disaster risk reduction. Capacity development at all levels of society, comprehensive knowledge management and the involvementof science (including behavioral science) in public awareness raising, media communication, behavior change, and education campaigns should be strengthened.

In education sector, The Ministry of Education (MoE) revised the General Science Subject for lower secondary school curriculum in 2006 and included the study on 'Earth and Space' with lessons on storms. The lower secondary life skills subject also covers flood, emergencies, earthquake, tsunami, landslides and fire. The revised upper secondary school curriculum includes a lesson titled 'Earthquake' in Grade 10 English and 'Earth Surface Process' in Grade 11 Geography. At the primary level, a chapter on Caution in Emergencies is included. A complementary reading material that contains information on 8 disasters is available as a self-study booklet for Grade 5, 6 and 7 students. General Studies Textbook (Level-2) with 'Earthquake', 'Storms', 'Tsunami', and 'Preparedness' topics and a story book 'Be prepared' are available for Non-Formal Education. Recently, DRR along with State/Regional Government and MOE has incorporated Do's and Don'ts on various natural hazards into the student exercise book provided by the Government.

Myanmar National Building Code (MNBC) is being developed to enhance the skills of persons engaged in construction sector. Myanmar Environmental Conservation Law (2013) can cover stipulation of the environmental quality standards, environmental conservation and management of urban environment, conservation of natural resources and cultural heritage. '***Women and Emergencies***' is one of the priority areas to strengthen systems, structures and practices to ensure women's and girls' rights to protection in emergencies and to ensure their participation in emergency preparedness, response and disaster and conflict risk reduction.

4. Summary

Synergies with the climate change and sustainable development requires collaboration and communication across the scientific disciplines and technical fields, and with all stakeholders including representatives of governmental institutions, communities of policy making, scientific and technical specialists and members of the communities at risk to guide scientific research, set research agendas, bridge the various gaps between risk assessments and risk perception by stakeholders, and support scientific education and training. The potential contribution of affected and vulnerable communities in generating research questions, and in performing research, collaboratively or independently, should be valued and facilitated. An overall understanding of disaster risk reduction in Myanmar concerning with S&T awareness, development and education will be given by this paper.

References

1. Centre for Excellence in Disaster Management and Humanitarian Assistance. Myanmar (BURMA) Disaster Management Reference Handbook, assessed in January 2017
2. DRR Task Force of Government 2009. Myanmar Action Plan on Disaster Risk Reduction (MAPDRR) 2009-2015.
3. Reid Basher, 2013. Science and Technology for Disaster Risk Reduction: A review of application and coordination needs.
4. Relief and Resettlement Department 2015. Risk Assessment Roadmap Myanmar.

Chapter 21

Restoration Opportunities Assessment Methodology (ROAM) for Landscape Stewardship from Natural Disasters: A Way Forward

R. Srivastava

Centre for Science and Technology of the Non Aligned and Other Developing Countries (NAM S&T Centre), New Delhi, India
E-mail: fe.rashmii@gmail.com

ABSTRACT

The geo-climatic conditions and socio-economic vulnerability of India ranks the country at 75th place on the World Risk Index of natural disaster in 2017 (UNU-EHS). This implies that lack of critical infrastructure and weak logistic chains will gradually drag the country towards the increased risk of extreme natural events. The time is now ripe to look beyond the planning aspects and step in for implementation stages. One such effort is based on the phenomenon of landscape restoration technique known as Restoration Opportunities Assessment Methodology (ROAM), which has been conceptualized by the International Union for Conservation of Nature and Natural Resources (IUCN) and World Resources Institute (WRI). It is a forward-looking and dynamic approach that incorporates powerful combination of stake-holder engagement ('best knowledge') and analysis of documented data ('best science') to identify and investigate Forest Landscape Restoration (FLR) opportunities applicable to the area in question in order to reinforce disaster risk reduction and landscape resilience. By integrating potential economic and carbon sequestration via cost-benefit modeling, it also encapsulates economic viability of the assessments to facilitate validation of strategic recommendations for preventing natural extreme events. Not limiting itself to field data estimates, it further opens the avenues for accomplishing the Bonn Challenge target to restore 150 million ha of land worldwide by 2020 and substantially contribute to national and international programmes viz., National Adaptation Programmes of Actions (NAPAs), United Nations Programme on Reducing Emissions from Deforestation and Forest Degradation (UN-REDD+),United Nations Convention to Combat Desertification (UNCCD), United Nations International Strategy for Disaster Reduction

(UNISDR) and United Nations Framework Convention on Climate Change (UNFCCC). Additionally, in order to comprehend the methodological framework of ROAM, its pilot applications conducted in Rwanda (2013) are reviewed and discussed in the paper. In due course, this robust tool has finally been initiated in the Indian states of Madhya Pradesh and Uttarakhand by IUCN and WRI team members, which are notably the two most disaster prone (flood, landslides, drought) regions of the country, thereby mitigating, improving and restoring the landscapes and livelihoods.

Keywords: *ROAM, FLR, Carbon sequestration, Landscape restoration, Bonn challenge, National and international programmes.*

1. Introduction

A resilient planet needs robust science for disaster risk reduction.

– Ms. Margareta Wahlström

Reducing disaster risk is everybody's business, and needs everyone's participation and investment – civil society, professional networks as well as municipal and national governments.

– Ban Ki-Moon,

United Nations Secretary-General,
on the occasion of the International Day for Disaster Reduction
("Curbing disaster risk," 2010)

We are witnessing today dramatic upheavals and sweeping climate extreme trends owing to both natural and anthropogenic in origin. Their consequence result into deterioration of human and environment ecosystems with serious and notable impacts on economies, food security and built infrastructures of affected regions. In recent past, it has been found that change in climate patterns is influencing global warming which is exacerbating the frequency of natural disasters (**Figure 21.1**) *viz.*, typhoons, cyclones, earthquakes, floods, drought, landslides, heat and cold waves *etc.* The percentage of occurrence of such extreme natural events across the world by disaster type is illustrated in **Figure 21.1** hierarchically, which represents that the flood (43 per cent) and storms (28 per cent) causes the most adverse impacts on human and built environment, whereas wildfires (4 per cent) and volcanic activity (2 per cent) are least responsible for the damages (UN/CRED, 2016).

Besides, the statistical reports of CRED and UNISDR (2016), reveals that most of the developing countries across the world witnesses 70-163 and 164-472 number of hydro-climatic meteorological disasters.

Over the years, the most cited questions being evoked time and now are how long-term weather patterns will influence society in a more profound way and how we should respond to it or, as rainfall intensifies and sea levels rise, how and when should vulnerable populations be re-located permanently out of harm's way? To help answer these queries, Ms. Margareta Wahlström, Representative of the Secretary General for the United Nations International Strategy for Disaster Reduction (UNISDR), said that the Scientific and Technological capacities and capabilities must be considered in its widest sense to facilitate the remedy for climate

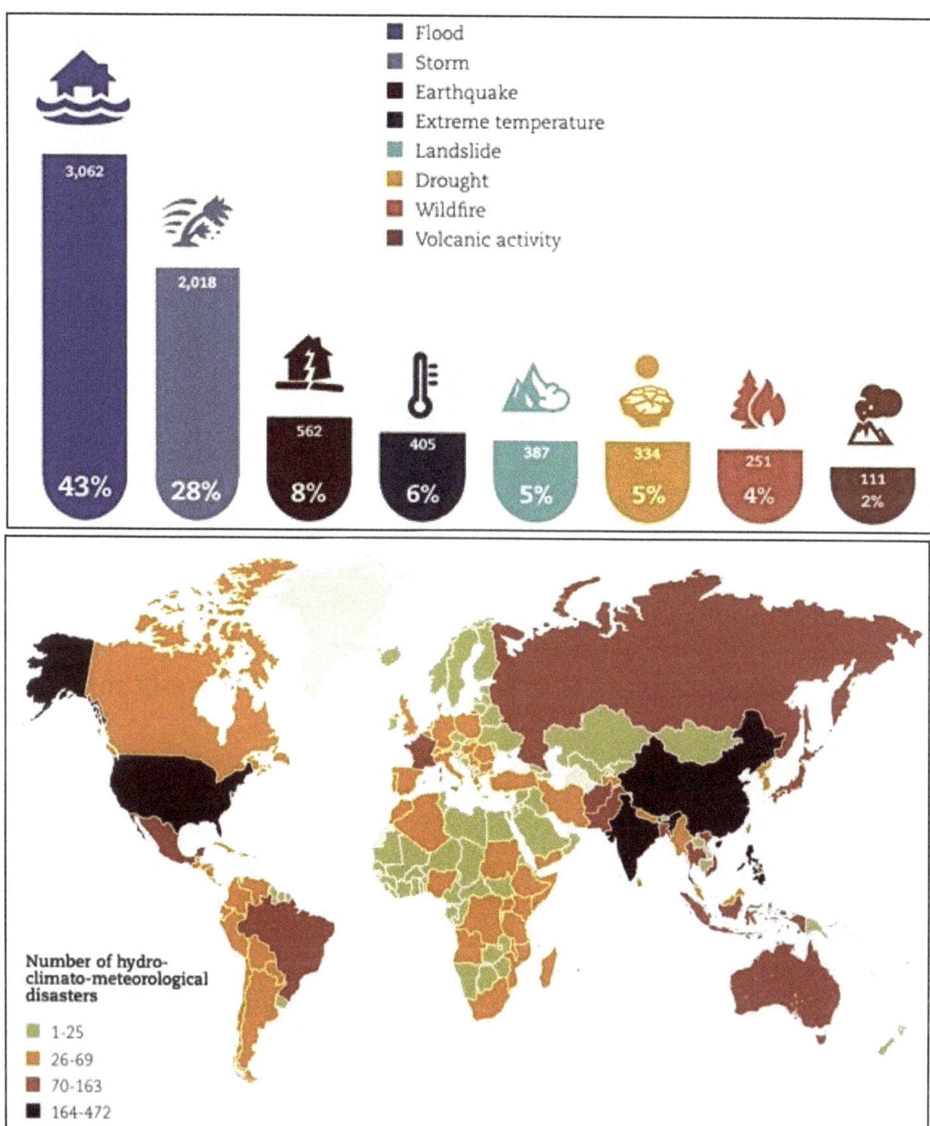

Figure 21.1: Percentage of Occurrence and Number of Weather Related Disasters Reported per Country from 1995 to 2015 (*Source:* CRED and UNISDR, 2016)

extremes across the globe. According to the available scientific estimates and the recent report of World Risk Index (UNU-EHS, 2017), it has been estimated that 171 countries of the world are at the verge of experiencing adverse impacts of climate extremes in the near future. The map created by them (**Figure 21.2**), demarcates various 'global hotspots' at higher disaster *risks* were located in Central America, Central and West Africa, Southeast Asia and Oceania; accumulation of very high

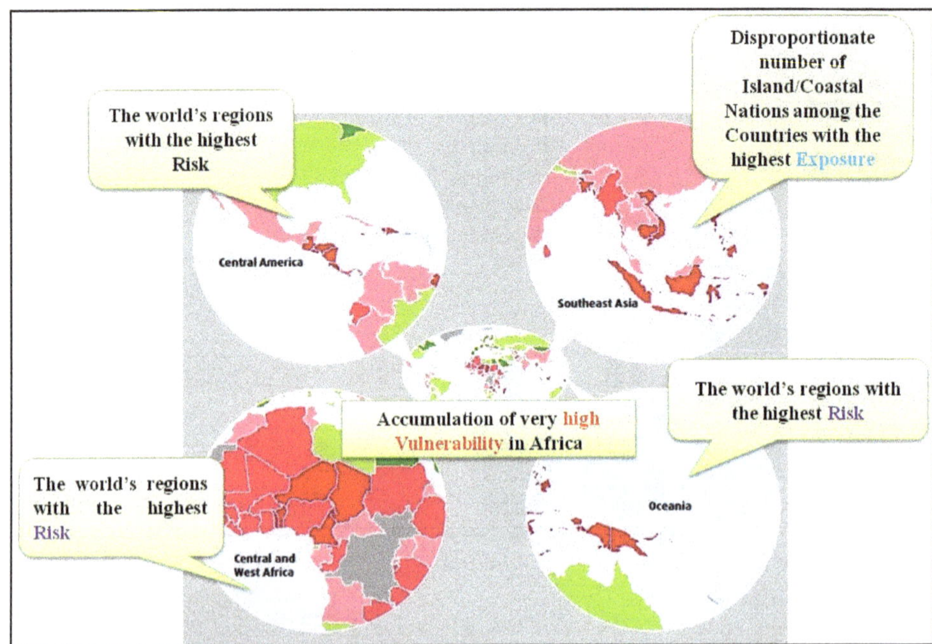

Figure 21.2: Global HOTSPOTS for Climate Extremes (*Source:* UNU-EHS, 2017).

vulnerability was examined in Africa; and the highest *exposure* to the number of disproportionate islands and coastal nations was observed in Southeast Asia (Wahlström M. 2013).

Additionally, global patterns of rainfall erositivity has resulted in unrecoverable soil erosion (**Figure 21.3**) across the different contours of the world which has disintegrated many ecosystems and food chains making the biosphere more susceptible to climate induced disasters (Panagos, 2017).

The datasets published by UN-University Institute for Environment and Human Security Source (UNU-EHS) has further investigated that the geo-climatic conditions and socio-economic vulnerability of India and ranks it at 75th place on the World Risk Index of natural disaster in 2017 with exposition, vulnerability and risk of 11.94, 58.62 and 7 per cent, respectively. The country mostly experiences 3100 – 5200 MJ mm ha^{-1} h^{-1} yr^{-1} of soil erosion, which is pushing the region more towards the risk of disaster vulnerabilities.

The frequency and intensity of natural disasters that the country has encountered in the past decade is rising and is taking a toll on people and economy. In 2017, nearly 40 per cent of the districts in India face the prospect of drought, while close to 25 per cent districts have had heavy rainfall of more than 100 mm in just a matter of hours. The year also saw five Indian states, Chandigarh, Mumbai, Bengaluru and Araria in Bihar, grappling with floods, with Assam facing the worst deluge in over a decade. Almost half of India, including food bowl states of Punjab, Haryana, Maharashtra and Madhya Pradesh, is facing drought. This implies that lack

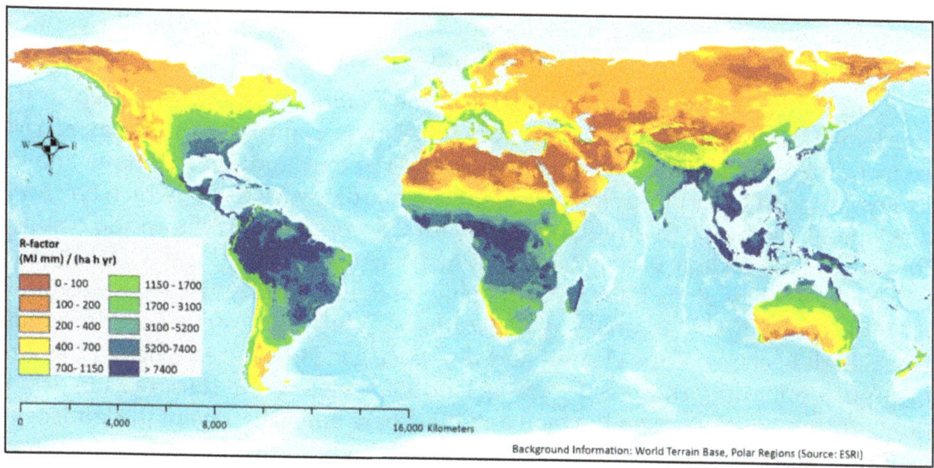

Figure 21.3: Global Rainfall Erosivity Assessment for Identifying and Analyzing Regions for Disaster Risk Reduction (*Source*: Panagos, 2017).

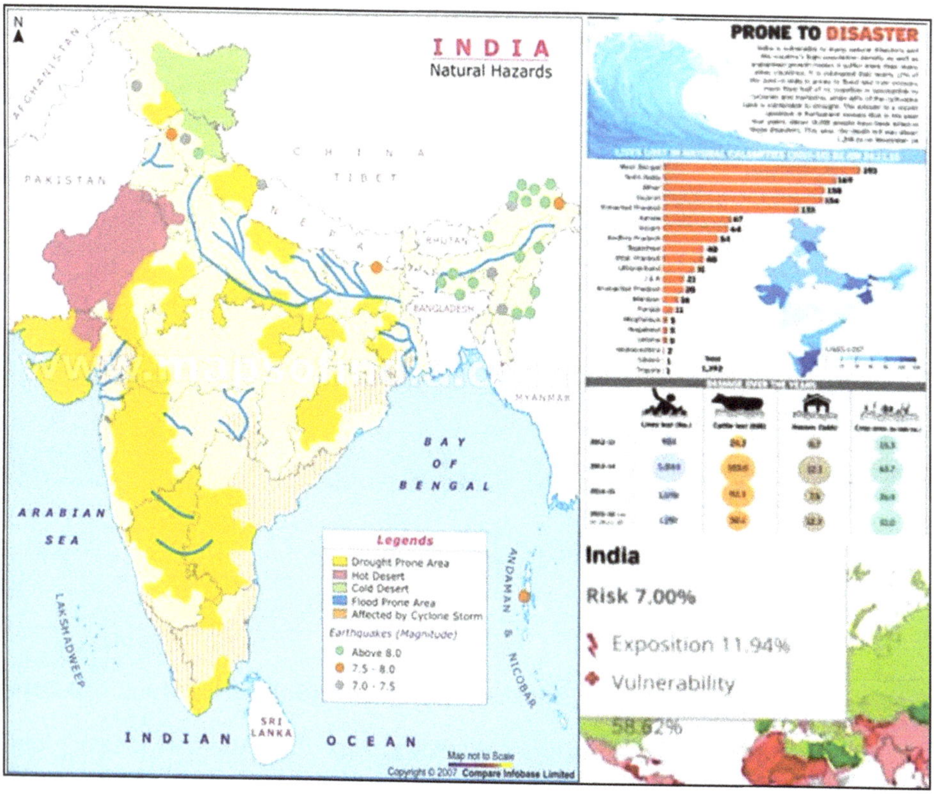

Figure 21.4. World Risk Index for India and Various Natural Disasters, India is Prone to (*Source:* Ministry of External Affairs and Times of India December 3, 2015; UNU-EHS-2017; and Map of India, 2016).

of critical infrastructure and weak logistic chains will gradually drag the country towards the extreme risk of extreme natural events.

The time is now ripe to look beyond the planning aspects and step in for implementation stages. One such problem-solving approach in disaster risk reduction is based on the phenomenon of landscape restoration technique known as Restoration Opportunities Assessment Methodology (ROAM), which has been conceptualized by the International Union for Conservation of Nature and Natural Resources (IUCN) and World Resources Institute (WRI). In 2011, WRI, IUCN and partners on behalf of the Global Partnership on Forest and Landscape Restoration (GPFLR) published *'Landscapes of Opportunity'* which indicated that more than two billion hectares of the cleared and degraded lands - an area twice to the size of China - offer opportunities for forest landscape restoration (**Figure 21.5**). It not only protects and conserves the man and biosphere but also prevent the landscape against natural jeopardy. These include 700 million hectares in Africa, 400 million ha in Asia, and 500 million ha in Latin America.

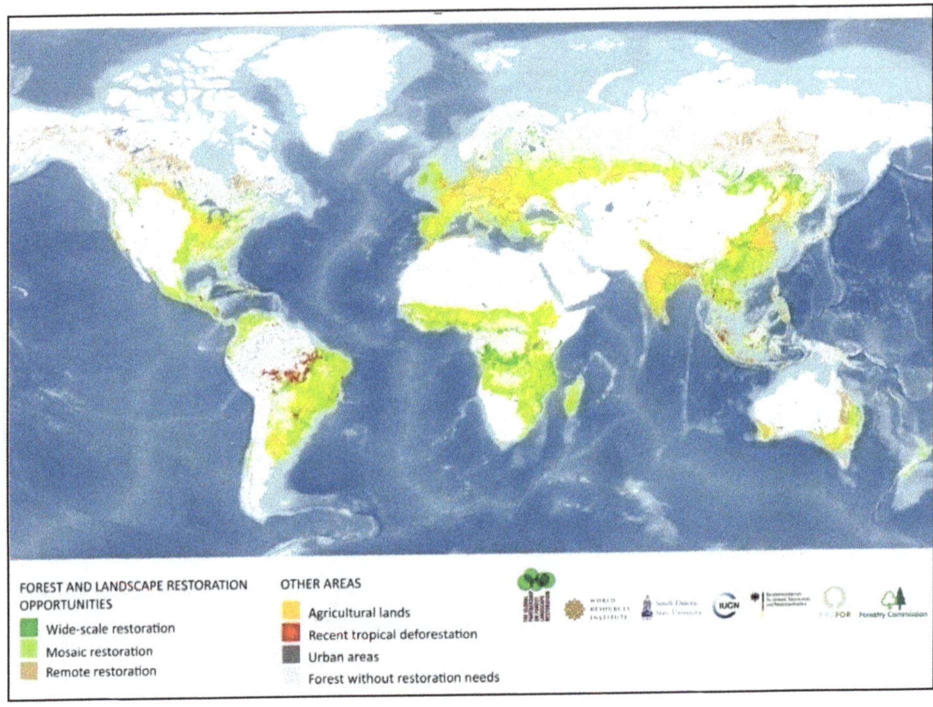

Figure 21.5: Global Potential for Landscape Restoration (*Source:* IUCN and WRI-2014).

The methodological framework of ROAM encapsulates this core idea of global potential for landscape restoration to make the planet more resilient to climate extremes. It is within these benchmarks, the initiative also encourages accomplishing sustainable goal for disaster risk reductions.

2. An Outline of ROAM: A Dynamic Approach

ROAM is a forward looking and affordable framework that incorporates powerful combination of stake-holder engagement ('best knowledge') and analysis of documented data ('best science') to identify and investigate Forest Landscape Restoration (FLR) opportunities applicable to the area in question in order to reinforce disaster risk reduction and landscape resilience. It is a stepwise and iterative application of a series of analyses (**Figure 21.6**) at a national or sub-national level, which is designed to help address the following common and essential questions for resilient landscape restoration:

1. Where is restoration socially, economically and ecologically feasible?
2. What is the total extent of restoration opportunities in the country/region?
3. Which types of restoration are feasible in different parts of the country?
4. What are the costs and benefits, including carbon storage, associated with different restoration strategies?
5. What policy, financial and social incentives exist or are needed to support restoration?
6. Which are the different concerned stakeholders with whom we need to engage?

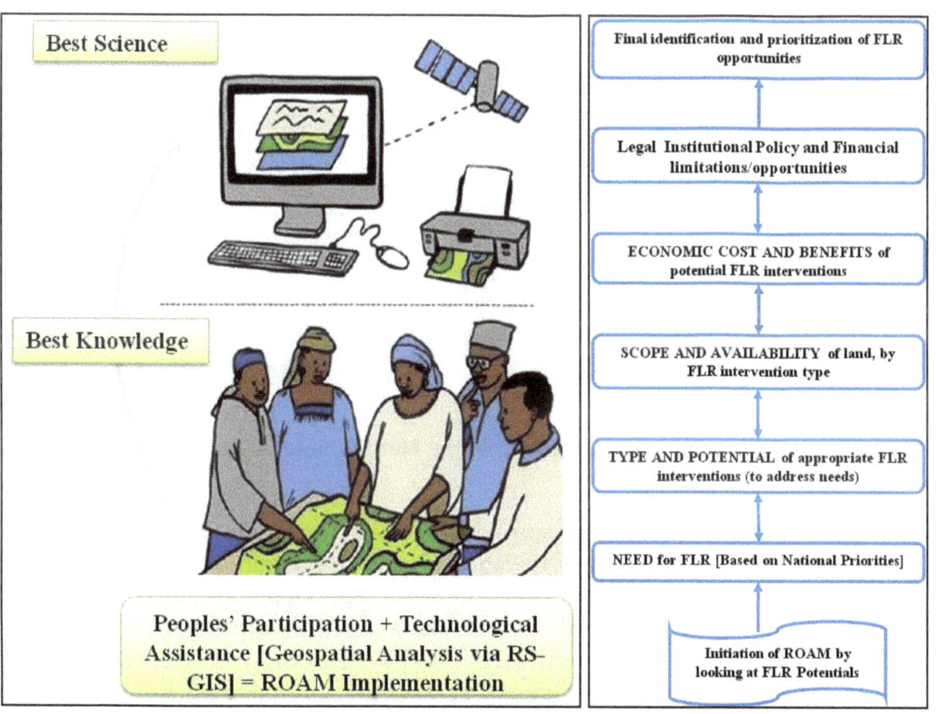

Figure 21.6: Schematic Representation of ROAM (*Source:* IUCN and WRI, 2014).

A National level assessment of ROAM typically requires 15-30 days of work by the assessment team, which is spread over a 2-3 month period. It's 'pilot' applications were conducted in Ghana, Mexico and Rwanda, each of these was tailored to provide specific analytical insights and policy recommendations; and is now being applied as the *model* to various regions of the world. By integrating potential economic and carbon sequestration via cost-benefit modeling, it also encapsulates economic viability of the assessments (**Figure 21.7**) to facilitate validation of strategic topographical recommendations for preventing natural extreme events. Nonetheless, this uncover existing capacity to improve disaster risk reduction and adaptation of ROAM has built the momentum towards the Bonn Challenge, a global commitment to start restoring 150 million ha of lost and degraded forests by 2020 (**Figure 21.8**). Incidentally, it also provides a basis to accomplish some of the most promising goals for national and international orientation *viz.*, National Adaptation Programme of Actions (NAPAs), United Nations Reducing Emissions from Deforestation (UN-REDD+), UN-Convention to Combat Desertification (CCD), UN-International Strategy for Disaster Reduction (ISDR), UN-Framework Convention on Climate Change (FCCC), Sendai Framework for Disaster Risk Reduction; and the most ambitious and holistic, Sustainable Development Goals (SDGs) of 2030.

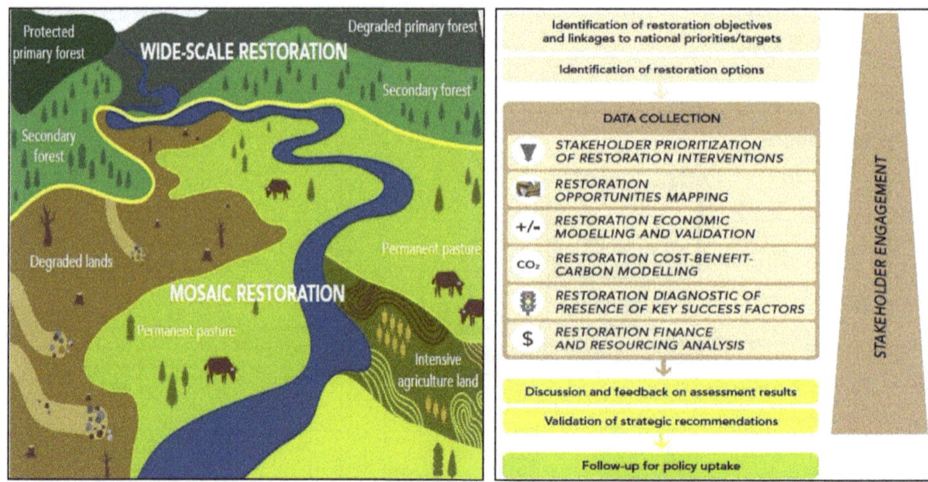

Figure 21.7: Schematic Representation of the Methodological Framework Adopted in ROAM.

In the year 2011, Rwanda made an ambitious pledge to the Bonn Challenge *i.e.* to restore 2 million hectares of land, establishing itself as a global leader in the restoration movement. Professionals from the Department of Forestry and Nature Conservation in Rwanda Natural Resources Authority (RNRA) worked in partnership with the experts from IUCN and WRI along with the other relevant governmental and non-governmental stakeholders to conduct ROAM in their country. This paper identifies priority areas for restoration in Rwanda (*as the case study*) that would reinstate the degraded and deforested land, in order to mitigate

Figure 21.8: Map Representing Adoption of ROAM in 23 different Geographies Across the World (*Source:* Messinger, 2015).

extreme natural events and help their citizens to foresee improved livelihood and landscape conditions.

3. Objectives

This chapter broadly engages with the following objectives:

- To present an overview of different components of ROAM for integrative planning in land use decisions in various topographies;
- To conceptualize the theoretical framework of ROAM specifically for disaster risk reductions; supported with the case study of Rwanda;
- To utilize this model as the roadmap for Madhya Pradesh and Uttarakhand (two of the most disaster prone regions of India, among others), and
- To maneuver the pathways for accomplishing the specific disaster risk related sub-targets of Sustainable Development Goals (SDGs) 9 (industry, innovation and infrastructure) and 11 (Sustainable Cities and Communities).

4. Methodology

In this study, I have tried to navigate a way forward (in conceptual terms) for precisely mitigating the natural extreme events from the prism of ROAM. For this purpose, I have conducted a meta-analysis by relying mostly on the theoretical framework of ROAM, conceived by IUCN and WRI, through research papers, stakeholders' reports, newspaper articles and online portals.

5. ROAM on the Ground with Specific Reference to Disaster Risk Reduction: A Case Study of Rwanda

5.1 Background Study

The Republic of Rwanda is a small, green mountainous, land-locked country situated in the east-central Africa (as represented in **Figure 21.9**). It's a densely populated developing country with 12,488,199 people as its inhabitants (NISR 2012). The current population density in Rwanda is 507 per Km^2 (1,312 people per mi^2) with total land area of 24,670 Km^2 (Worldometers 2018). The average annual growth and the GDP of the country is 2.6 per centand1, 302 billion Rwandan Francs, respectively (MINIRENA, 2013). The country mostly endures moderate climate as it is situated at an altitude above 1,000m and witnesses two rainy seasons (*i.e.* from Feb-June and Sept.-Dec).

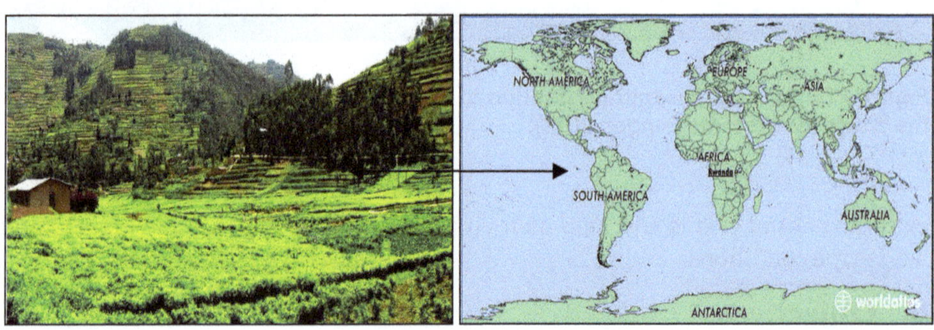

Figure 21.9: Location of Rwanda (*Source*: Harvey, 2015; and World Atlas, 2015).

Despite having the land-locked geography, the region still has adequate water resources *viz.*, lakes and rivers, along with the few exploitable land resources (Habiyambere, 2009). The country has 11 big cities and shares borders with Burundi, Tanzania, the Democratic Republic of the Congo (**Figure 21.10**). The country's population thrives on land, flora, fauna and water to meet their livelihood security, which is primarily based on agriculture and agroforestry (70 per cent), energy and timber (16 per cent) production. The Food and Agriculture Organisation of the United Nations (FAO, 2013) estimated that more than 40 per cent of the cultivated land in Rwanda possess a risk of severe erosion, *i.e.* 10 tons of soil loss per ha per year, which contaminates the rivers and streams that are not adequately protected, and make the region more susceptible to climate-induced disasters. Additionally, the various topographies of the country also witnesses other natural threats *viz.*,

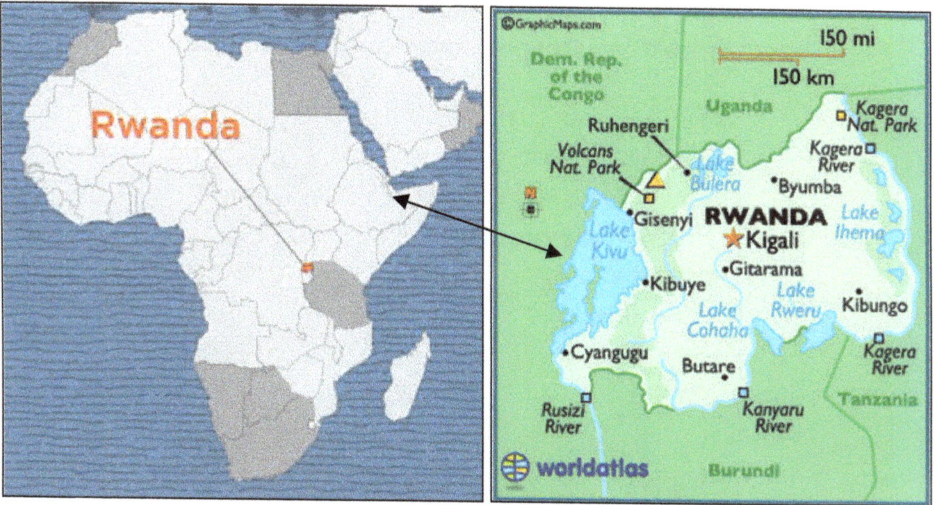

Figure 21.10: Site Map of Rwanda (*Source:* Harvey, 2015 and WorldAtlas, 2015).

floods, volcanic eruption, torrential rains and landslides due to increasingly global climatic changes. Notably, the country status reports clearly states that the region might face on-land poverty primarily due to soil erosion and drought, if stewardship against natural hazards are not implemented effectively (Habiyambere, 2009).

5.2 ROAM Implementation

It is within these complex environmental settings, the country has envisioned to achieve sustainable development and become a middle-income country by 2020 (**Figure 21.11**).This vision was encapsulated by a national consultative process, which started in 1998, to clearly define and device the future goals of Rwanda (Ministry of Lands, Environment, Forestry, Water and Mines, 2003). National governments and regional partners of the country are working in collaboration to achieve Vision 2020 goals. For instance, Belgian Development Agency and Rwandan Natural Resource Authority (RNRA) are working in close partnership to minimize deforestation and poverty by strategically planning and managing the existing woodlots and reforesting degraded and sensitive land (Belgian Development Agency, 2012). Besides, the implementation of ROAM is another path-breaking approach initiated by the country, which aims to restore degraded land by increasing the forest cover to 40 per cent and providing 100 per cent access to clean water by decreasing soil erosion and increasing water filtration services in the forests. Further, it also intends to practice agro-forestry on existing agricultural land for reducing drought prone factors in different contours of the country.

The Department of Forestry and Nature Conservation professionals of Rwanda Natural Resources Authority (RNRA) in partnership with IUCN and WRI experts worked in close association to apply the methodological framework of ROAM appropriately for disaster risk reductions apart from other land use decisions. The integrative planning among key stakeholders and communities provided

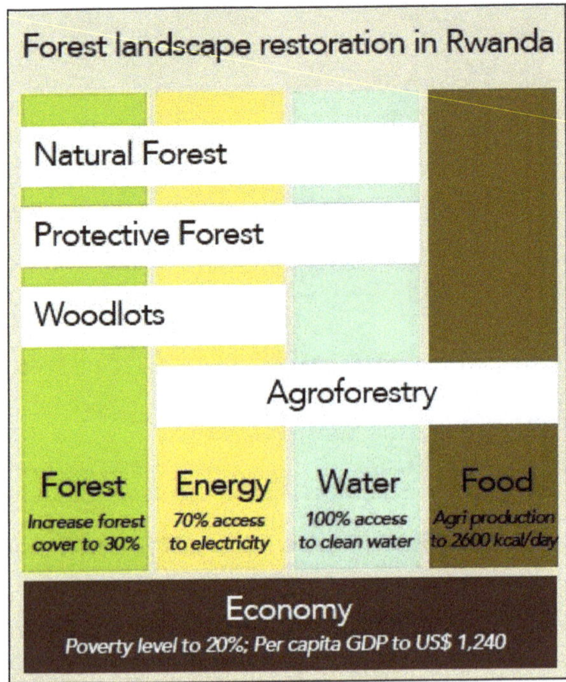

Figure 21.11: Potential Contributions of FLR Interventions to National Development Targets in Rwanda (*Source:* MINIRENA, 2013).

the framework for identifying high risk areas, vital ecosystems and vulnerable populations, to build resilient country against potential natural disasters (**Figure 21.12**).The integrative planning takes into the account the cumulative impacts on reducing disaster vulnerabilities by mapping the areas and landscapes with the most urgent restoration needs, *i.e.* where benefits are most immediate and where success is more likely.

To define output and scope of the assessment, three FLR priority regions, namely, forest land, agricultural land and protective land and buffers were outlined (see **Table 21.1**), wherein forest and protective land and buffers deciphered on the mechanisms to examine risk profile regions for natural hazard reductions. Across the region, it clearly illustrated the geophysical and climatic settings to delineate the significance of potential opportunities for improved integration in national disaster risk reductions platforms.

Thereafter, depending upon the social and topographical dynamics, the country was categorized into *seven* zones (**Table 21.2** and **Figure 21.13**) so as to bring about the detailed output for addressing cluster of natural threats effectively. The protective forest interventions were further categorized into following five prospective areas to prevent extreme weather events (MINIRENA, 2013)

1. Protective forest on ridge tops - slopes greater than 30 degrees/55 per cent incline;

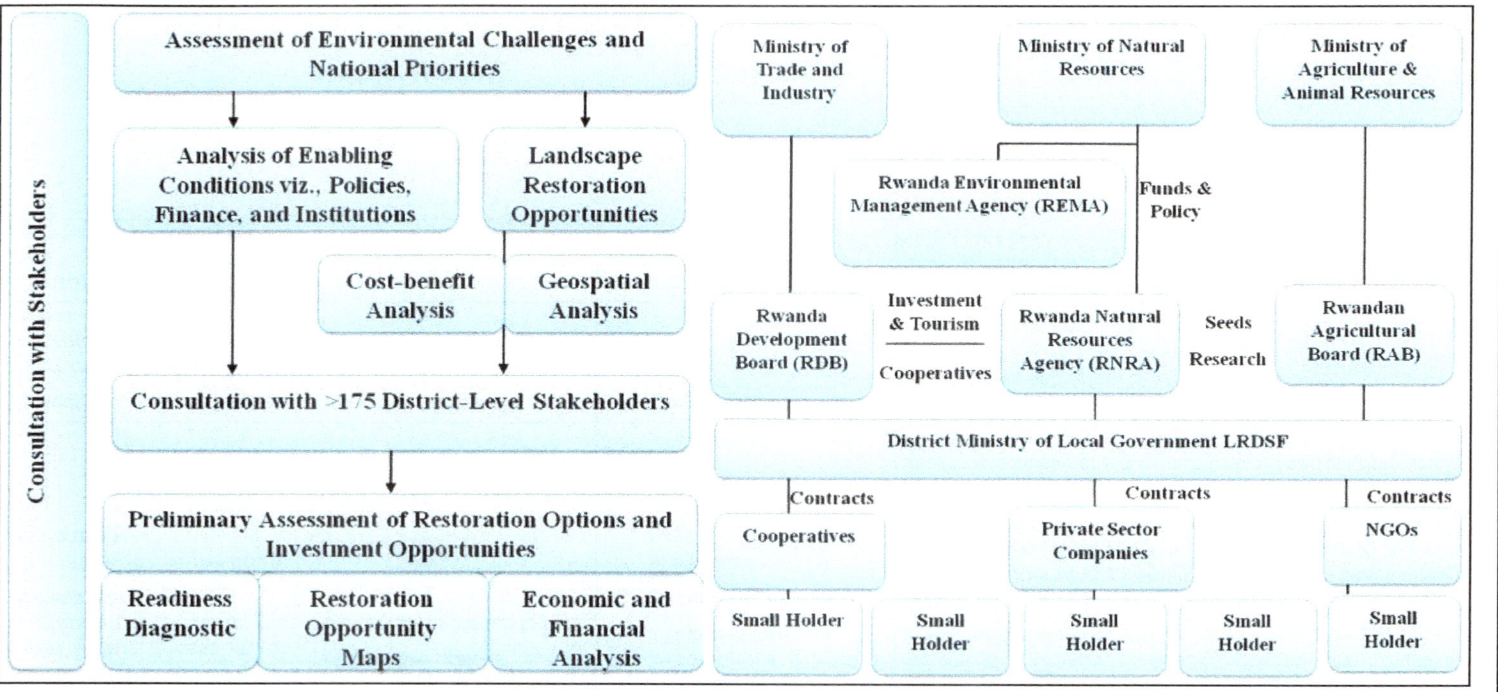

Figure 21.12: Flowchart Representing the ROAM Methodology Adopted by the Key Stakeholders in Rwanda (*Source:* MINIRENA, 2013).

Table 21.1: Key Landscape Restoration Priority Regions, with Respect to Natural Extreme Events, outlined by the Stakeholders in Rwanda (*Source:* Adopted by IUCN and WRI Report, 2014)

Land Use	*Land Sub-Type*	*General Category of FLR Option*	*Description*
Forest Land *[land where forest is, or is planned to become the dominant land use]* **Suitable for wide-scale Restoration*	If the land is without trees, there are two options If the Land is degraded forests:	1. Planted Forests and Woodlots, and Natural Regeneration 2. Natural Regeneration 3. Silviculture	*Planting of trees on formerly forested land.* Native species or exotics and for various purposes, fuel wood, timber, building, poles, fruit production, *etc.* *Natural regeneration of formerly forested land.* Often the site is highly degraded and no longer able to fulfill its past function- e.g. agriculture. ***If the site is heavily degraded*** and no longer has seed sources, some planting will probably be required. *Enhancement of existing forests and wood-lands of diminished quality and stocking, e.g.* by reducing fire and grazing, and by liberation thinning, enrichment planting *etc.*
Protective Land and Buffers *[land that is vulnerable to, or critical in safeguarding against, catastrophic events]*	If degraded Mangrove:	Mangrove Restoration	Establishment or enhancement of Mangroves along coastal areas and in estuaries.
**Suitable for Mangrove Restoration, Watershed protection and erosion control*	If other Protective Land or buffer:	Watershed Protection and Erosion Control	**Establishment and Enhancement of Forests on very steep sloping land, along water courses, in areas that naturally flood and around critical water bodies**.

2. Protective forest on ridge tops - slopes between 12-30 degrees/20-55 per cent incline;
3. Planting native tree species to create 20m buffers of non-forested river courses;
4. Replacing existing eucalyptus plants with indigenous tree species within 20m of river courses, and
5. Planting native species as buffers within 50 m of wetlands

Table 21.2: Stratification in *Seven* Sub-areas of the Country for ROAM Findings to Address Natural Hazards (*Source:* IUCN and WRI, 2014).

Sl.No.	*Strata*	*Features (Based on existing data sets)*
1.	Lake Kivu Shore	High population in certain districts, ***high erosion vulnerability***, ***high rainfall***, presence of key sectors that's impact or rely on natural resources (export crops, hydro-energy, mining, tourism)
2.	Central Plateau	***Highly degrades soils***, elevated poverty rates, significant fuelwood deficit
3.	Amayaga	Lowland, ***elevated drought risk***, structured land reform, presence of key natural resource dependant sectors
4.	Eastern Ridge and Plateau	***Highly degraded soils***, elevated poverty rates, high population pressure
5.	Eastern Dryland Savanna	Lowland, ***elevated drought risk***, good soil, high evapo-transpiration
6.	Buberaka Highland	High population, significant fuel wood deficit, acidic soils, low temperature
7.	Volcano and High Plains	Basic soil, high fertility, high population, presence of key natural resource dependant sectors (tourism, export crops)

Such recognition from intensive and inclusive stakeholders deliberations at multi-sectoral level outlaid elaborative deliberations and analysis, which facilitated in identifying policies that can encourage government officials, experts, and landowners to adopt restorative land uses practices, which may prevent the topography of the country to undergo extreme weather conditions.

5.3 Key Findings

The following results of the ROAM analysis of the country was obtained for mainstreaming disaster risk reductions:

- ✰ The potential areas for protective forest interventions were mostly located in the North (7,000 hectares *i.e.,* 2 per cent of the province area), South (14,000 hectares *i.e.,* 2 per cent of the province area) and Western (15,000 hectares *i.e.,* 3 per cent of the total province area) provinces
- ✰ Overall, 42,000 hectares of land, *i.e.* 2 per cent of Rwanda's total area, were the most suitable areas reforesting ridge tops, in order to stabilize and prevent soil erosion

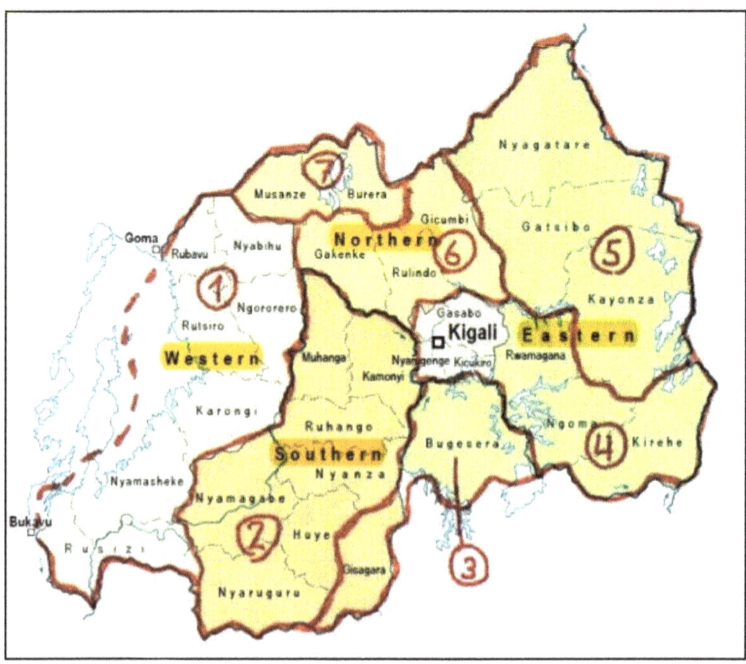

Figure 21.13: Stratification in *Seven* Sub-areas of the Country for ROAM Findings to Address Natural Hazards (*Source:* IUCN and WRI, 2014).

- The total area for watershed management across the country estimated to be relatively small. However, the extensive networks of river/streams and wetlands provide significant potential areas to strengthen adjacent soil and mitigating flow of sediments into the nearby water bodies, throughout the country
- To build 20m buffer zone of native tree species in non-forested regions and/or to substitute the eucalyptus stands (23,000 hectares *i.e.*, 1 per cent of the total area of Rwanda) by planting indigenous tree species;
- Nationally, approximately 57,000 hectares *i.e.*, 2 per cent of the total area of Rwanda mostly in the East- 3 per cent and South- 4 per cent) for creating potential 50m buffer zones by implanting native tree species along the perimeters of wetlands
- The economic and financial analysis depicted that the returns on investments for the private landowners was not positive and they need to invest significant amount of labor and materials while altering deforested land into protective forests, and in return they might receive few income, if any marketable benefits occur. However, this transition lead to the ample amount of environmental benefits *viz.*, protection of flood plains, wetlands, rivers, lakes and other vulnerable contours from degradation
- In combination with agro-forestry and restored/improved forests on ridge tops with steep slopes, approximately 40,000 hectares of land effectively

addressed sedimentation problems and resulted into better soil restoration and erosion control

- ☆ Approaches like, plantation of timber and non-native fruit tree species, was employed to improve the capacity of protective forest without preceding revenue generation
- ☆ Restoration of degraded natural forests, particularly inside the national parks, was addressed by planting *Pinus* tree in a 100m buffer region across the periphery. This additionally enabled in biodiversity conservation and encouraged eco-tourism
- ☆ It enabled in comprehending the importance of better connectivity within the ecological corridors between forests like, Gishwati, Nikita and Nyungwe, of national priority, and
- ☆ Restoration of dry land forest in Easter province of the country was urged to prevent landscapes from drought or flooding due to increase in vulnerabilities of drier and flatter areas by climatic changes

Thus, the so obtained key observations from ROAM framework provided number of concrete actions for the country to analyze and prevent hazard risks and vulnerabilities. They not only assisted in modifying the frequency of natural weather patterns, but it also served in realizing the protection of ecosystems to prevent natural barriers that can moderate the impact of hazards and safeguard communities.

6. Weaving the Conceptual Framework of ROAM for the Two Most Disaster Prone Cities in India

6.1 Background

In the historic speech about India's '***tryst with destiny***' on the midnight of 14th August 1947 (eve of India's Independence), Jawahar lal Nehru - the first Prime Minister of India – expressed his views in how he envision the newly independent India with rest of the countries and said- "Those dreams are for India, but they are also for the world, for all the nations and peoples are too closely knit together today for any one of them to imagine that it can live apart. Peace has been said indivisible; so is freedom, so is prosperity now, and so also is disaster in this one world that can no longer be split into isolated fragments" (The Guardian, 2007). It is this landmark utterance that captured the dreams for India and its interaction with the rest of the world, the Prime Minister not only articulated for peace and economic prosperity but also put across his opinion to ward off the planet against the natural apocalypse.

If one closely examines the international disaster database report published by Centre for Research on the Epidemiology of Disasters (CRED), then it can be deciphered that 2017 was the second most costliest year (after 2011), in terms of both human and economic losses, caused due to natural extreme patterns, globally. It also reflects that 90 per cent of deaths were due to meteorological and climatological disasters, and almost 60 per cent of people were impacted by floods and 85 per cent were affected due to economic damages (**Figure 21.14**) caused by the

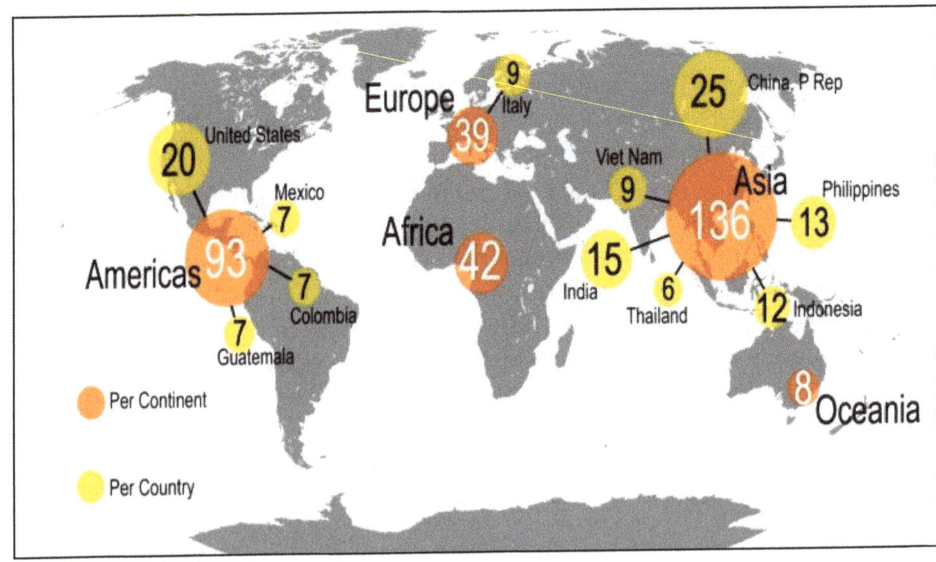

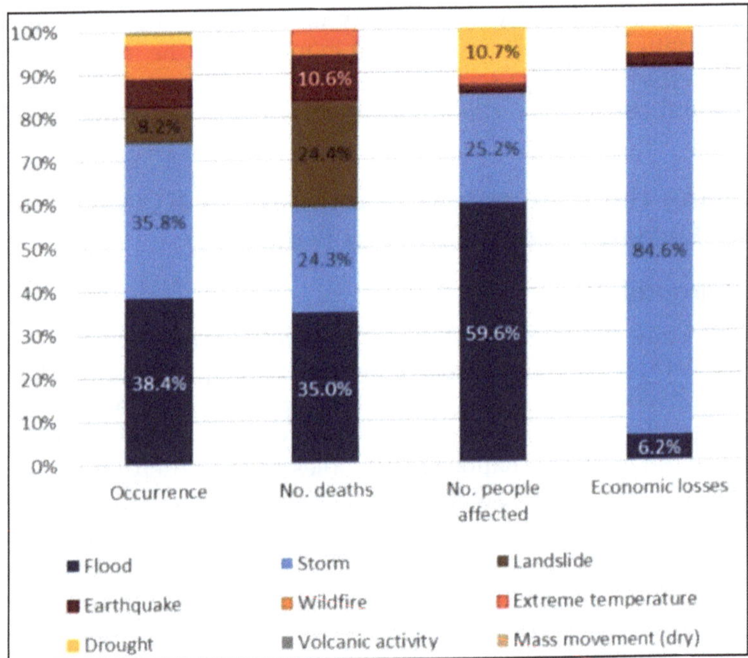

Figure 21.14: The Weather-Related Disasters Witnessed by different Countries and the Share of different Vulnerabilities by Disaster Type Across the World for 2017 (*Source:* Wallemacq, 2018).

storms (Wallemacq, 2018). In midst of these statistical data analysed for the world, the findings for India are also unappreciative, which reveals that the impact of natural extreme events on humans' were highest in 2017. It has resulted into early

2,300 deaths and has adversely impacted lives of 22.5 million of people (affected predominantly by floods and storms).

The country which ranked at the 62nd place among the emerging economies on an Inclusive Development Index – India - is becoming the fastest mushrooming regions in the world, along with the interdisciplinary interrogation of the growth in the country that promises to yield new theoretical and practical insights. At such a stage, it is imperative for environment manager's *viz.*, Scientists, Policymakers, Academicians, Civil Societies *etc.*, to collect and analyse the relevant data as an indicator of progress for disaster risk reduction.

6.2 In the Tale of Two Indian Cities – Uttarakhand and Madhya Pradesh

As increase in climate changes are threatening large swathes of the globe, two of the Indian cities, Madhya Pradesh (MP) and Uttarakhand (UK), also experiences unprecedented natural hazards *viz.*, drought, floods, lightning, fire, torrential rains, hail storms, land and mud slide *etc.* While the data on ranking different states of India is yet to be published, still it can be noted from the records that these two Indian states had experienced severe climate induced disasters.

In Madhya Pradesh (second most largest state of India, geographically), out of 50 districts 32 districts in last 30 years have witnessed flood, and around 7 districts were highly affected from drought (State Disaster Management Plan – Madhya Pradesh, 2015). Among all the natural hazards, floods are most frequent and devastating in the MP, with some of the extreme flood destructions in the history *viz.*, in 1982, 1983, 1984, 1986, 1992, 1994, 1996, 1997, 2003, 2005, 2006, 2012, 2016, 2018 (recent year). Further, it has been observed that the state experiences more than 80 per cent of the annual rainfall for a period of 3 months (short monsoon), which leads to a heavy siltation, flash floods and poor discharge of flood waters.

On the other hand, Uttarakhand (the major Hindu pilgrimage site) being mountainous terrain covered by dense forest, is easily susceptible to extreme natural hazards *viz.*, floods, land and mud slide, torrential rain (major cloud burst in 2013), deforestation and forest fires. The state experiences frequent denudation due to melting of glaciers, violation of environmental laws, building of hydro-electricity plants and excessive tourism.

Notably, both these states have their unique and profound socio-economic, diverse culture and ecological importance. Therefore, it is imperative for these regions of the country to develop robust integrative planning and resilient infrastructure against natural calamities. Keeping this in purview, ROAM multi-sectoral approaches can be used as a model for improved cumulative response on reducing natural disaster vulnerabilities. It will pilot its initiatives in seven different countries (including India) namely, Kenya, Vietnam, Peru, Uganda, Mexico and El Salvador. This approach has bridged the gap between land-based mitigation and adaptation, and has facilitated the decision makers to maximise the co-benefits economically and socially against the natural catastrophes.

For this purpose, IUCN in partnership with the G.B. Pant National Institute of Himalayan Environment and Sustainable Development (GBPNIHESD),

Uttarakhand - for technical inputs – and Federal Ministry for the Environment, Nature Conservation and Nuclear Safety (BMU), Germany – for donation - has initiated the pilot project of ROAM in Uttarakhand. The state was selected because it has the tremendous prospects for achieving land-based mitigation and adaptation against disaster vulnerabilities.

In March 2017, IUCN and GBPNIHESD have begun the assessment in the entire state, with the intensive stewardship in two districts namely, Garhwal (commonly known as Pauri Garhwal) and Pithoragarh. Following that recently, on 8th March 2018, the state-level stakeholders consultative meeting was held in which, representatives from various relevant institutions and departments took initiatives to optimise the implementation of ***'best-bet'*** regions and approaches for landscape restoration and protect the state from natural calamities (**Figure 21.15**). The validation seminar included experts from diverse fields namely, the Department of Forest, Agriculture, Sericulture and Social Welfare, Watershed and Fisheries Management Directorates, Uttarakhand Livestock Development Board, State Horticulture Mission, GBPNIHESD, Forest Survey of India (FSI), Forest Research Institute (FRI), Indian Institute of Remote Sensing (IIRS), Indian Council of Agricultural Research-Indian Institute of Soil and Water Conservation (ICAR-IISWC), Wildlife Institute of India (WII), Community Representatives, Uttarakhand Rural Development Department, Council for Higher Education Accreditation (CHEA), Uttaranchal Youth and Rural Development Centre, State Medicinal Plants Board (SMPB), Uttarakhnad State Rural Livelihood Mission, Maiti Sangathan, State Biodiversity Board, WRI, Indian Council of Forestry Research and Education (ICFRE), Indira Gandhi National Forest Academy (IGNFA), Uttarakhnad State Council for Science and Technology, UNDP, Disaster Mitigation and Management Centre and IUCN (Proceedings of the ROAM Validation Workshop for Uttarakhand - India, 2018). The event thus deliberated upon, the broad scale flexible and cost-effective restoration diagnostics for the inclusive deployment of national, regional and local land-based mitigation and adaptation strategies to address food, energy and water security along with the disaster oriented vulnerabilities.

Moving forward to the other state of India – Madhya Pradesh (MP), ROAM adaptation began on 22nd July, 2016 in Bhopal by organizing the two day workshop with an objective to bring together all the stakeholders' of the state at a common platform for integrative and collaborative planning and assessments (**Figure 21.16**). It has to be noted here that the ROAM application in MP is at the nascent stage and is still undergoing series of consultative dialogues among different participatory authorities. Therefore, it would be too premature to mention here, about the deliberations, suggestions and findings of ROAM in this state. However, the roundtable held in 2016 can be outlined to comprehend the initiatives taken by the state to minimize climate-induced vulnerabilities. The meeting included stakeholders' from political and civil societies of MP *viz.*, Members of Legislative Assembly (MLA), Department of Culture, WRI, Forest Department, State Election Commissioner, EPCO – Institute of Environmental Studies and IUCN, wherein the key highlights included the hands-on stewardship opportunities for land-use measures, environment impact assessment, monitoring changes in natural

Figure 21.15: State-level Stakeholders' Consultative Meeting on ROAM for Uttarakhand (*Source:* IUCN, 2018).

hazards and promoting R&D for improved integration and mitigation of challenges witnessed by the state in previous decades. In this context, Mr. R. Parasuram (State Election Commissioner of MP) said, *"Livelihoods should form an integral part of the landscape restoration approach taken in Madhya Pradesh. Our strategy must enhance the communities' resilience to climate change"* (Anon, 2018).

Figure 21.16: State-Level Stakeholders' Meeting on ROAM for Madhya Pradesh (*Source:* Anon, 2018).

6.3 Probable Outputs

As both the projects of ROAM in India are still being carried out and are expected to get completed by the end of this year. Therefore, the published findings and recommendations are unavailable. Nonetheless, they have demarcated the following expected outputs from ROAM to grapple with the natural catastrophic events:

- Empowering '***best-bets***' regions for optimizing rational implementation of land-use dynamics at national, sub-national and local levels
- Provide road-map for disaster risk reduction for broad range of environmental initiatives *viz.*, carbon sequestration, biophysical degradation *etc.*, along with the economic plan that prioritize these initiatives
- By achieving this, it would directly as well as indirectly facilitate the implementation of several existing national and international commitments *viz.*, National Adaptation Programme of Actions (NAPAs), Sustainable Development Goals (*specific SDGs for disaster management are: 9 - Industry, Innovation and Infrastructure and 11 - Sustainable Cities and*

Communities), Convention on Biological Diversity (CBD) Aichi Target 15, UNFCCC, Bonn Challenge, REDD+, Rio+20, Conference of Parties (COP) 21, UNCCD, ISDR, Sendai Framework for Disaster Risk Reduction and UNISDR

- It will provide important opportunities for mainstreaming R&D and science and technological (S&T) interventions for preventing natural hazards, and will strengthen the environmental components of disaster reduction
- Strengthen the establishment of national, regional and local integrated disaster preparedness and early warning systems, to reduce the adverse impacts on the people and their livelihood caused by natural hazards
- Such decentralized engagement would create positive impact on the livelihood practices of the communities *viz.*, farmers, fisher folk, grazers *etc.*, and would explain and inform them on the climate induced changes
- Would aid in mobilizing commitments and coordinating investments among broad range of actors *viz.*, Public-Private Partnership (PPP), Corporate Social Responsibility (CSR), sponsorships from international institutions *viz.*, World Bank, International Monetary Fund (IMF), European Investment Bank *etc.*, to fund and prevent land fragmentation across the region;
- It will prioritize the implementation of forestry based ecosystem rehabilitation *viz.*, watershed management (RAMSAR sites), flood plain and protective natural forest restoration, buffer zones, ecological corridors, low-lying land and very small remnants of natural forest management, littoral and riverine stewardship *etc.*, with relatively lower costs;
- Would help different environment managers/decision makers in organizing outreach programmes, relating to the significance of natural ecosystems and land-use planning for effective flood-risk mitigation and rainwater harvesting
- Propose and mainstream equal gender participation in natural resource management to scale up the sustainable land management along with the improved food, water and energy security
- Would identify and promote aggregation groups that can finance *via* Small and Medium-scale Enterprises (SMEs) or cooperatives for disaster mitigation measures;
- Would build capacities in disaster risk reduction, by incorporating three important components namely, human resources, institutional development and financial resources, to increase awareness and maximize various opportunities for engaging in joint efforts
- Create and support institutions to extend farm credits to smallholders for enhancing their resilience and adaptive capacities, with an interest in investing in value added processing and plantation forestry, expansion of woodlots and marketing of tree and forest products

- ☆ Restoration of degraded forest contours (one of the most over looked opportunities for economic and social well-being), particularly under the reserved categories *viz.*, national parks, sacred groves, sanctuaries, small forest remnants *etc.*, will improve biodiversity conservation and will encourage eco-tourism, and
- ☆ Would act as the blue print for policy makers, planners, general public and scientists - together with the existing guidelines, at national/district level(s) to better comprehend their local climatic systems and will help them to formulate future governance related to disaster vulnerabilities, appropriately.

Thus, to accomplish the aforementioned third objective of this paper, it can be affirmed that this integrative tool is enabling enriched content-specific knowledge and understandings for integrative land-use planning and management, and can be employed as the yardsticks for addressing and evaluating the cluster of natural threats in these two cities of India.

7. Conclusion

As the world is currently ushering towards the fulfillment of 2030 agenda, it can be illustrated that the benchmarks of ROAM with its multi-sectoral dynamics, can assist in accomplishing specific targets corroborating to natural disasters of SDGs and other national and international commitments, along with the area-specific guidelines and cost-effective long-term way out to the communities (who are the first line of experts and respond to these crisis in an organic and dynamic manner) and environmental managers for disaster risk reduction.

We all are now interconnected, which makes all of us equitable in not only sharing the natural resources but it also, makes us more responsible to think about the rest of the planet more coherently in anything we do. Nowadays, global forces are more proactive and engaging, and are emanating from every conceivable direction. The international system of this 21st century, with its networked collaborations, will help this mushrooming world to promote and maintain a period of cooperative coexistence in any specific region and across the globe. In this context, Ban Ki-Moon, former Secretary-General of United Nations on the occasion of the International Day for Disaster Reduction (*"Curbing disaster risk," 2010*) remarked, *"Reducing disaster risk is everybody's business, and needs everyone's participation and investment – civil society, professional networks as well as municipal and national governments"*, So to conclude (*notwithstanding with its few limitations*), the ROAM initiative is one of those collaborative mechanism that can maximize more deliberate and integrative engagements to protect lives and human well-being against extreme disaster vulnerabilities.

References

1. Anon, 2018. Release: WRI India Commits to Implementing Landscape Restoration in Madhya Pradesh, Focusing on Climate and Communities | WRI India. [online] Available at: https://bit.ly/2LZFDDY (Accessed 7 Aug. 2018).

2. Belgian Development Agency 2012. Support program to the reforestation in 9 districts of the northern and western provinces of Rwanda.
3. CRED and UNSIDR 2016. Weather Related Disasters 1995-2015. [online] Available at: https://bit.ly/21CHVDW (Accessed 3rd Feb. 2018).
4. FAO 2015. Status of the World Soil Resources. [online] Available at: https://bit.ly/2iV3YNN (Accessed 13th Jan. 2018).
5. Habiyambere, T., Mahundaza, J., Mpambara, A., Mulisa, A., Nyakurama, R., Ochola, W. O., *et al.*, 2009. Rwanda State of Environment and Outlook. Kigali: REMA.
6. Harvey, M. 2015. In Deep: A Brief Introduction to Rwanda. [Blog] *Butterflies and Robinson.* Available at https://bit.ly/2T89LEC (Accessed 13th Feb. 2018).
7. IUCN and WRI 2014. A guide to the Restoration Opportunities Assessment Methodology (ROAM). [online] Available at: https://bit.ly/2rzPGEE (Accessed 11th Nov. 2017).
8. IUCN 2018. Uttarakhand state validates restoration opportunities and priorities identified using ROAM framework. [online] Available at: https://bit.ly/2AMtz3G (Accessed 6th Aug. 2018).
9. Messinger, J. 2015. Global Restoration Initiative [online]. Available at: https://bit.ly/1Q2EkQh (Accessed 17th January 2018).
10. MINIRENA 2013. Five year strategic plan for the environment and natural resources sector, 2014-2018. Kigali.
11. Mukushema Adrie et. al. 2014. The ministry of natural resources – Rwanda. International Union for Conservation of Nature and Natural Resources (IUCN) and World Resources Institute (WRI).
12. NISR 2012. Population and housing census. National Institute of Statistics (NIS) of Rwanda. Kigali.
13. Panagos, P. et. al. 2017. Global rainfall erosivity assessment based on high-temporal resolution rainfall records. *Scientific Reports* 7: 4175.
14. Proceedings of the ROAM Validation Workshop for Uttarakhand, India (2018). [online] Dehradun, IUCN. Available at: https://bit.ly/2MvkRse (Accessed 7 Aug. 2018).
15. Republic of Rwanda, Ministry of Lands, Environment, Forestry, Water and Mines. (2003). National Environmental Policy. Kigali.
16. The Guardian 2007. A Tryst with Destiny. [online] Available at: https://bit.ly/2b1E9py (Accessed 6 Aug. 2018).
17. Wallemacq, P. 2018. Natural disasters in 2017: Lower mortality, higher cost. [online] Brussels: UCL and USAID. Available at: http://file:///C:/Users/st-1/Downloads/CredCrunch50.pdf (Accessed 6 Aug. 2018).
18. Wahlström M. 2013. Using Science for Disaster Risk Reduction. Report of the UNISDR Scientific and Technical Advisory Group. http://www.interacademies.org/24913/Using-Science-in-Disaster-Risk-Reduction

19. World Atlas 2015. Where is Rwanda ? [online] Available at: https://bit.ly/2I2DASK (Accessed 13th Feb. 2018).

20. World Atlas 2015. [online]. Available at https://bit.ly/2mRyxm7 (Accessed 20th January 2018).

21. Worldometers 2018. Rwanda Population (2018) - Worldometers. [online]. Available at: https://bit.ly/2QGJVWK (Accessed 2nd Feb. 2018).

Chapter 22

Location-based Disaster Notification System

T.T. Zan and S. Phyu

University of Computer Studies, Yangon, Myanmar
E-mail: thuthuzan@ucsy.edu.mm, sabaiphyu@ucsy.edu.mm

ABSTRACT

Natural disaster cannot be prevented, but its impacts can be reduced or rescued. Adequate prior disaster information can save a significant number of lives in some affected situations such as train derailments and gas-related fire damage. Myanmar has notification services such as broadcasting from media and sending messages from the operators. It also broadcasts to all people for advertising and other events such as entertainments, product promotions and so on. In this system, disaster notification is sent to registered mobile users who are in the disaster prone area. The system finds whether a mobile is within a defined disaster area using its GPS coordinates. The system architecture is built for sending notifications to mobile devices in disaster prone area. Then, multicast notification service is utilized to send notifications only to those who need to receive them. The effective information and broadcasting of potential disaster to mass people going to affect, might be useful preventive measure to secure lives on large scale.

Keywords: *Android notification techniques, Location based service (LBS), Push technology, Global positioning system (GPS), Google cloud messaging (GCM), Circle property.*

1. Introduction

Everyone who is in IT field says *'today is the age of three things: Cloud Computing, Internet and Mobile'*. This is true because there is no doubt that businesses can reap huge benefits from them. In mobile technology, the traditional ***'Pull Technology'*** has become outdated because it requires that users know a prior notification where and when to look for data (Rundle *et al.*, 2015). It suffers from transmission latency and duplicate data traffic. Then, a more accurate description of a process is appeared and it is called automated Pull Technology (*Push Technology*). However, all delivered information is wanted by the users.

In this paper, ***Push Technology*** is used that allow the server to multicast the warning messages for incoming disasters according to the client regions. This system focuses on Android because it is used by majority of smartphone devices in the world.

The objectives of this chapter are:

1. To propose Mobile Region Check (MCR) procedure for determining mobile region
2. To propose hybrid update procedure that will help to get the current positions of moving mobile at client side and reduces the server update cost greatly.

The advantages of this system are the following, first message can send to all without using operator and second, it remains as notification which users can get without missing any alert. Third, the system is mostly work on server that is reducing and managing the information overload as a result mobile user's save battery life of phones.

2. Related Works

There are a number of papers that describe about timely disaster prevention. Disaster Management Center of the University of Wisconsin said the term Disaster management can be described as *"The range of activities designed to maintain control over disaster and emergency situations and to provide a framework for helping at-risk persons to avoid or recover from the impact of the disaster"*. Most papers are focusing on their nations and discuss not only prevention but also evacuation for mobile users.

A mobile application was developed to help in the efficient provision of rescue and relief to disaster affected areas in India (Barapatre *et al.*, 2014). This application is used for sending the location wherever disaster has taken place. It is also used by user who can provide help in affected areas. It works with two buttons, '**I NEED HELP**' and '**I WANT TO HELP**'. So this makes an interaction between the victim who is facing disaster and the volunteer who desire to help the victim. GPS based coordination in between victim and volunteer is problematic sometime because lack of details on Google map for navigation (Barapatre *et al.*, 2014).

Saravana Kumar and Veeramani proposed their own algorithm '***Extended Polygon Match Algorithm using Quadtree***' to find whether a mobile is within a defined polygon shaped area using their GPS coordinates. Their objectives are to build a prototype system using android software to send alerts to mobile devices within the defined geographical area and to fix the search area and accurate location is easy using GPS. The advantage of their system is that the disaster target region is perfectly contained. But the other regions also contain when taking the corner points of the polygon in map. They have planned to implement this concept in all the mobiles which is equipped with GPS (Kumar S. *et al.*, 2014).

Amit Gosavil, Vishnu proposed to notify the user located in possible disaster zone with visual and audio disaster warning and evacuation guideline combine with nearest location of shelter or safe zone on the map of the application. This system helps out to both normal and blind people to reach to the nearest safe place prior to disaster. However, lack of details on Google map of developing countries is the main challenge of their work. They have to implement an application for rescue and relief operation with better server side application to totally automate the system of detecting disaster prone area as future work (Gosavi A. *et al.*, 2014).

Yavuz Selim,Yilmaz Bahadir, Ismail Aydin Murat Demirbas evaluate arrival times to elaborate how GCM performs (timing performance of GCM), Poisson distribution to the number of devices per time and conducted *chi-square* goodness-of-fit test on their models (Selim Y. *et al.*, 2014). They point out GCM servers on client device does not by itself guarantee a timely message arrival.

Harminder Singh, Dr. Sudesh Kumar, Harpreet Kaur explains how GCM service and location service can be combined to develop a new kind of service (Singh H. *et al.*, 2015). They point out the message delivery using GCM is unpredictable, the devices may respond to user commands with different delay. The device must have stable internet connection.

3. Notification Type

This section explains notification type for android using push and Google cloud messaging. The worldwide smart phone market grew 13 per cent year over year in 2015, with 341.5 million shipments. According to data from the International Data Corporation (IDC), **android** dominated the smart phone market with a share of 82.8 per cent (Nzegbuna P. 2015). In Myanmar, android mobile penetration claims to reach 95 per cent at the end of 2015 (Techinasia, 2015). Android is used by the majority of smartphone devices in the world. It is popular because it is free and has an easy-to-use user interface. The android operating system has received several updates since it was introduced in 2008 and it's currently at Ver.5.

It has been estimated that 71 per cent of all app developers in the world develop apps for the Android market, which keeps expanding daily.

4. Location Based Services

Location Based Services (LBSs) are services offered through a mobile phone and take into account the device's geographical location.

LBSs typically provide information or entertainment. LBS largely depend on the mobile user's location.These services can be classified into two types: Pull and Push. In a Pull type, the user has to actively request for information. In a Push type of service, the user receives information from the service provider without requesting it at that instant (Technopedia, 2015).

5. Push Technology

Push technology is a style of Internet-based communication where the request for a given transaction is initiated by the publisher or central server. Push is also known as 'Webcasting', 'Netcasting' or 'Pointcasting' (Käpylä T. *et al.*, 2003). Push notification is a kind of notification that has been sent out to a device by a central server. Push has more capacity than pull technology such as immediacy, efficiency, reduced latency, longer battery life, shorter learning curve.

Push notifications can be sent according to the services from windows store app that use Windows Push Notification Service (WNS). iPhone and iPad apps are based on the Apple Push Notification Service (APNS). Push notifications are sent to android apps by using the Google Cloud Messaging (GCM) service (GCM Overview, 2016).

6. Firebase Cloud Messaging (FCM)

Firebase Cloud Messaging (FCM) is a cross-platform messaging solution that lets you reliably deliver messages at no cost. It inherits the reliable Google Cloud Messaging (GCM) infrastructure. GCM is used for sending messages from cloud to android device. FCM also allows the developer to send data from third party server to their applications running on android devices. It handles queuing of messages and delivers to the target application running on the target device. It supports Push notifications and can payload up to 4 Kb.

7. Global Positioning System

Global Positioning System (GPS) uses a constellation of between 24 and 32 earth orbit satellites that transmit precise radio signals, which allow GPS receivers to determine their current location, the time, and their velocity. GPS relies on a constellation of at least 24 satellites to provide location, speed and direction information to its users. It works by using a technique called trilateration combined with atomic clocks in the satellites in order to accurately determine the correct location (Chaudhari A. *et al.*, 2015).

8. Proposed Approach

This Paper are integrated by four major components- (1) Defining disaster region based on disaster types; (2) Determining registered mobiles which are in imminent disaster area by Mobile Region Check (MRC) procedure; (3) A proposed Hybrid Update Algorithm is applied to the client side that will help to reduce update cost; (4) An architecture is built to send notification message the target mobiles in the defined region.

9. Marking Disaster Area

This module is used to mark targeted region in map, generate circle region for marked area. This system uses circle property that has the set of points in a plane that are a fixed distance from a given point, called the center.

For example, this system use region with circle mark when disaster earthquake become around two miles in a region. **Figure 22.1** shows marking disaster affected area and service area for registered mobiles.

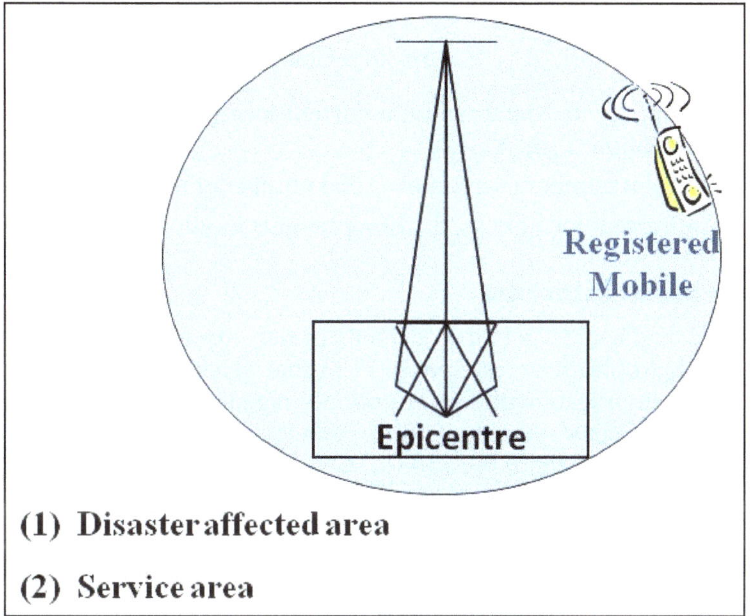

Figure 22.1: Marking Disaster Area by Circle.

10. Determining Registered Mobile's Area

The system determines whether registered mobiles are in service area or not so that this system compares distance between mobile device and epicenter with circle's radius. If the radius of circle is greater than or equal to the distance, the mobile is inside the service area. The following Mobile Region Check (MCR) procedure is used to decide registered mobile region.

voidMbCircle (float m_latitude, float m_longitude, c_latitude, c_longitude, double radius){double distance = sqrt((double)(c_latitude - m_latitude) * (c_latitude - m_latitude) + (c_longitude - m_longitude) * (c_longitude- m_longitude))

if (distance <= radius) printf ("\n Given point is inside the circle"); else printf ("\n Given point is outside the circle");}

11. Hybrid Update Algorithm

Algorithm: Hybrid Location Update

Input: Database of mobile locations contains the locations of registered mobile with time

Output: current registered mobile location

1. i = 0; dis_threshold; time_threshold; time_scheduler;
2. read the current location (L_{xi}, L_{yi}, t_i) and previous location($L_{xi-1}, L_{yi-1}, t_{i-1}$) of registered mobile location

3. if the time_scheduler > time_threshold and

   $\sqrt{(L_{xi}-L_{xi}-1)^2 + (L_{yi}-L_{yi}-1)^2}$ > dis_threshold

   3.1. Update the database with current location (L_{xi-1}, Lyi_{-1},t_{i-1}) = current location (L_{xi},L_{yi},t_i),i+1.

   3.2. Total number of update = Total number of update+1;

4. Else current location (L_{xi},L_{yi},ti) = previous location(L_{xi-1},L_{yi-1},t_{i-1}),i=i+1;

12. System Architecture

Android application gets the current position through GPS from the user's mobile phone. Application registers to FCM that generates register ID, then the application communicate with server not only register but also send the latitude and longitude of user's current position. For registered users who are located in disaster area, the server sends message to FCM. Afterwards, application fetches the message from FCM. The architecture of the system is given in **Figure 22.2.**

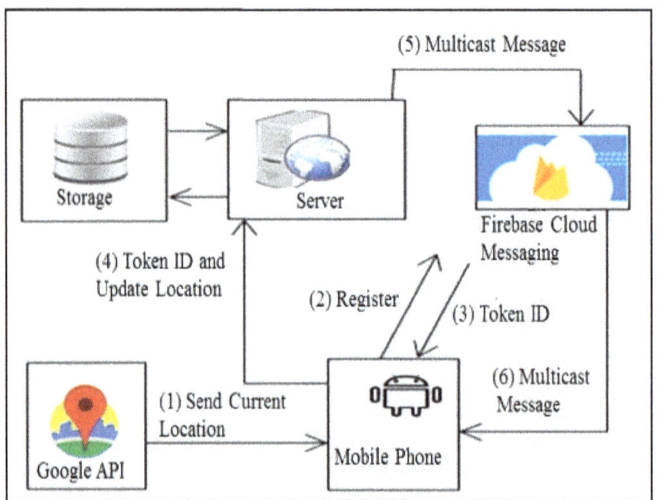

Figure 22.2: Communication between Mobile Phone, FCM and Server.

13. Implementation

When opening ***NotiApp***, Green colored App Bar and Tag Bar can be seen to represent fine Weather condition. At home fragment will appear Today's, Yesterday's and Tomorrow's weather condition of current location. Besides, it is possible to search by one city in search box; all three weather conditions will be appeared in **Figure 22.3a**.

By pressing News tab, it will show some photos with names of disaster causing in around Myanmar which is shown in **Figure 22.3b**. Supply tab contains help button that is placed to make phone call to WNI (Myanmar Office) and then to make contacts to local organizations from WNI is shown in **Figure 22.3c**. When the disaster

Figure 22.3: NotiApp, Showing (a) Gomepage (b) News (c) Supply Tabs.

occurs, the required notification is sent either Myanmar font or English font *via* FCM and registered mobiles will receive it and save as message is shown in **Figure 22.4**.

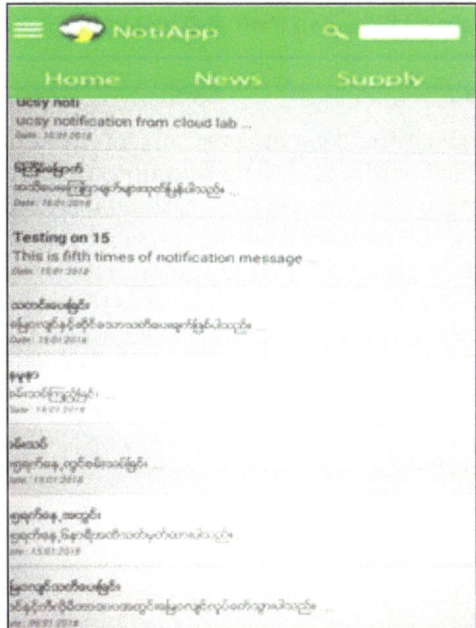

Figure 22.4: Notifications Saved as Messages.

14. Simulation Result

The simulation considers an experimental result with threshold value for proposed Hybrid Update algorithm. These values inserted in the local database of the moving object. Then compute the distance and if the distance > = a specific threshold, an update occur.

Figure 22.5 represents the actual and expected path through 9 minute at threshold = 2 miles.This result shows that central database needs to be update with actual location only five times at point a, b, c, d and e. Since the distance is greater than the value of threshold instead of updating the database every time.

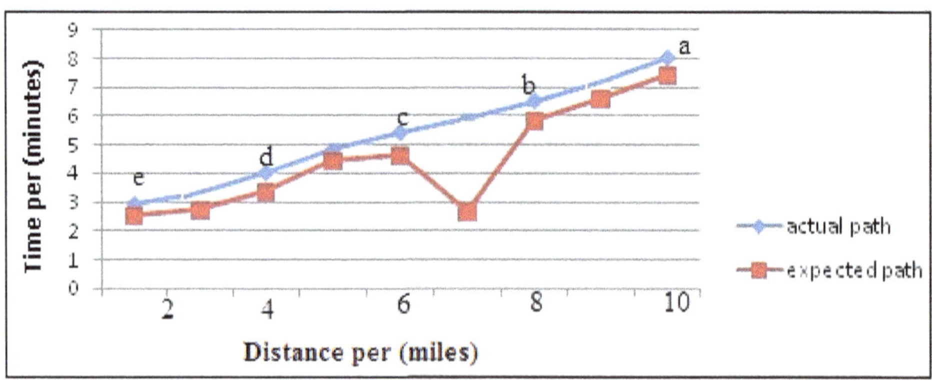

Figure 22.5: The Actual and Expected Path at (Threshold = 2 miles).

14. Conclusions

This system is aimed by notification especially for Myanmar weather conditions that are too bad and facing difficulties. Notifications or alerts will be delivered to users who are in the disaster prone area (*e.g.* earthquake, cyclone). The notification is sent as timely delivered message from server. FCM supports to access message that retain the queue in its server during offline. This system has to do further research to define disaster region for cyclone. It needs to ensure batches of messaging operations are committed atomically *i.e.* data synchronization between client and server. Moreover, this system will do authentication that confirm the message is only sent by unadulterated server. This system will take not to be clients whoreceive the same message multiple times and not to message expire before being accepted during time-to-live period.

References

1. Rundle, Huffington M. 2015. Future of technology: a white paper. London, United Kingdom
2. Disaster Management Centre at UW Wisconsin

   https://dmc.engr.wisc.edu/ and https://15projects.com/projects/the-need-for-a-disaster-management-centre-698606/, November 15th, 2014.
3. Barapatre H. 2014. Disaster management using android technology. *Journal of Research in Information Technology*, 2: 171-183

4. Kumar S., Veeramani 2014. GPS location alert system. *IOSR Journal of Computer Engineering* 16: 36-42.

5. Gosavi A., Vishnu S. S. 2014. Disaster alert and notification system via android mobile phone by using Google map. *International Journal of Emerging Technology and Advanced Engineering*, 4: 150-156.

6. Selim Y., Bahadir Y., Demirbas I.A.M. 2014. Google Cloud Messaging (GCM): An Evaluation. IEEE Global Communications Conference, 8-12 Dec. 2014, Austin, Texas, USA.

   https://ieeexplore.ieee.org/document/7037233

7. Singh H., Kumar S., Kaur H. 2015. Location based system using Google Cloud Messaging. Proceedings of National Conference on Innovative Trends in Computer Science Engineering (ITCSE-2015) held at BRCMCET, Bahal on 4th April 2015.

   https://www.ijrra.net/NCITCSE2015/NCITCSE53.pdf

8. Nzegbuna P. 2015. IT market analyst, PCs Notebooks, and Servers, International Data Corporation (IDC), West Africa, October, 2015. https://www.idc.com/in

9. Techinasia, 2015. https://www.techinasia.com, 2015.

10. Technopedia, 2015. https://www.techopedia.com, 2015

11. Käpylä T., Niemi I., Lehtola A. 2003. Towards an Accessible Web by Applying PUSH Technology. VTT Information Technology, PO-Box 1201, FIN-02044 VTT, Finland

12. http://www.vtt.fi/tte/

13. Google Cloud Messaging: Overview. 2016. https://developers.google.com/cloud-messaging/gcm, November 23, 2016.

14. Chaudhari A., Bohr S., Arma H., Dhupadale A. 2014. GPS/GSM enabled person tracking system. *International Journal of Innovative Research in Science, Engineering and Technology*, 4: 981-986.

Chapter 23

Enabling Rapid Disaster Response Using Artificial Intelligence and Social Media

M. Imran, F. Alam, F. Ofli and M. Aupetit

Qatar Computing Research Institute,
Hamad Bin Khalifa University, Doha, Qatar
E-mail: mimran@hbku.edu.qa

ABSTRACT

Disasters and emergencies often bring unanticipated situations for the members of public and formal response organizations. During such times, access to rapidly changing information plays an important role to understand the situation as it unfolds. However, information scarcity, especially in the early hours, hinders this task and ultimately delays response operations. Social media platforms such as Twitter and Facebook are gaining attention of both formal response organizations and affected individuals.People use social media to share a variety of information online including requests for help, and other urgent needs such as food, water, and shelter. This online information can be useful for humanitarian organizations if processed timely. This work presents a number of Artificial Intelligence (AI) technologies developed at the Qatar Computing Research Institute (QCRI) to aid disaster response and management. These technologies include machine learning platforms to automatically process textual messages (e.g. tweets) into humanitarian categories such as reports of needs, injured or dead people, and to process imagery data. The platforms work in real-time to collect and analyze data on social media to help humanitarian organization gain situational awareness and extract actionable information. These technologies were deployed during the Hurricane Maria and in this paperwe presentour findings and insights from this real-world natural disaster.

***Keywords**: Disaster response, Social media, Artificial intelligence.*

1. Introduction

According to The Economist, meteorological disasters caused by extreme weather including cyclones, blizzards, heat waves, hurricanes, droughts are increasing in the 21st century (Economist, 2017). While more people are being suffered by these disasters, the number of fatalities is actually decreasing as response to these disasters improves. However, the devastation in terms of human-lives and economic damage caused by these disasters is still huge. For instance, the recent Hurricane Harvey of 2017 is being estimated as the second-costliest natural disaster in U.S. history after Hurricane Katrina in August 2005 (Wikipedia, 2005 and 2017a). Disasters and emergencies bring uncertain situations in which formal responders and members of the affected communities struggle to find useful information, in particular first-hand and trustworthy information, to improve decision-making and to launch relief operations. For instance, the United Nations Office for the Coordination of Humanitarian Affairs (UNOCHA) seeks to gain situational awareness in the first 48 to 72 hours after a sudden-onset disaster to understand urgent needs of the affected population (*e.g.* food, water, shelter, medical assistance), disaster severity, and local government capacity to respond, among other factors (Whipkey and Verity, 2015). Gathering such information at the sudden-onset of a disaster, especially in the first 24 hours is nearly impossible, as it requires human experts and crisis severity assessors to visit the disaster areas, interview the affected people, and observe their urgent needs.

Access to rapid information as the crisis situation unfolds is a challenging task, particularly in the early hours of a disaster event when no other information sources are available (Imran *et al.*, 2015). To bridge this information scarcity gap, humanitarian relief organizations have recently started acknowledging the role of social media platforms such as Twitter and Facebook as vital information sources for disaster response and management (Whipkey and Verity, 2015). The widespread use of social media sites during disasters and emergencies provides convenient ways to share and consume public information as fast and easy as never before (Hughes and Palen, 2009; Vieweg *et al.*, 2010). People use web and mobile technologies to share different types of information including textual and multimedia content such as images and videos (Nguyen *et al.*, 2017). If accessed and processed properly, this information could be useful for a variety of disaster response operations (Imran *et al.*, 2015).

Despite the fact that information available on social media is timely and potentially useful, making-sense of it is challenging due to a number of reasons. Among others, the time-critical analysis of social media streams requires processing millions of messages arriving at high-velocity, real-time parsing of brief and informal content, handling information overload, determining information credibility, and prioritizing useful information for stakeholders (Boersma *et al.*, 2017; Castillo 2016; Imran *et al.*, 2015).

This work presents a number of Artificial Intelligence (AI) Technologies developed by the ***Qatar Computing Research Institute* (*QCRI*)** towards addressing the challenges in the use of social media for disaster response and to help relief organizations. On these lines, one of QCRI's flagship AI technologies is '***Artificial Intelligence for Digital Response* (*AIDR*)**' (Imran *et al.*, 2014). AIDR is a system conceived and developed at QCRI to harness information from real-time tweets that emerge from an area struck by a natural disaster to help coordinate relief activities.

The AIDR system combines human and machine intelligence to categorize crisis-related messages during the sudden-onset of natural or man-made disasters. The United Nations Office for the Coordination of Humanitarian Affairs (UN OCHA) now routinely uses AIDR for the coordination of humanitarian affairs as well as by many other emergency departments in the world. The system has been used during several major disasters such as the 2015 Nepal Earthquake, the 2014 typhoon Hagupit and the 2013 typhoon Hayan by a number of humanitarian organizations including UN OCHA and UNICEF.

In addition to textual messages, multimedia content such as images and videos on social media can be valuable for response organizations in a number of ways (Daly and Thom, 2016). For example, images can be used to determine damage severity done by the disaster (Ofli *et al.*, 2016). For this purpose, the AIDR system, which was originally designed to process textual content in real-time, has been extended to process imagery data posted on social media (Alam *et al.*, 2017). In order to make sense of imagery content on social media and to use it to aid relief operations, a number of challenges have to be addressed. Among others, these challenges include identifying and removing irrelevant and duplicate images, classifying images into different humanitarian categories such as injured people scenes, damage scenes, and shelter location scenes and so on. As a first step, currently we focuses on the task of disaster damage severity assessment. That is, we aim to use social media images to measure the extent of damage to critical infrastructure such as buildings, bridges and roads. Specifically, we have developed Deep Neural Networks-based models to assess the severity of damage shown in an image in three levels: **SEVERE** damage, **MILD** damage, and **NO** damage (Nguyen *et al.*, 2017).

Employing AI techniques to process potentially noisy social media data helps reduce noise. However, even after an intelligent filtering of noisy content, the remaining data, which is presumably useful, can still consist of tens of thousands of messages. Manual analysis of these thousands of messages remains a challenging endeavor for decision-makers, as it creates an information overload issue during an ongoing disaster*i.e.* new data arrives before human-assessors have finished analyzing the old data. To overcome this information overload issue, we have developed an interactive dashboard that helps crisis responders quickly sift through thousands of messages (Aupetit and Imran, 2017). The dashboard extracts critical entities (mentions of people, places, and organizations) from the data and allows users to apply multiple complex filtering criteria to find useful data nuggets.

To determine the effectiveness of all these technologies, we deployed all these systems during the 2017 Hurricane Maria (Wikipedia 2017b). We use AIDR machine learning classifiers to automatically classify Twitter messages into humanitarian categories and use images posted during the hurricane to assess the severity of damage. Furthermore, we use our interactive dashboard technology to find useful information during the event. This work presents the results of our analysis performed on the data collected from Twitter during Hurricane Maria.

The rest of the paper is organized as follows. The next section describes the artificial intelligence technologies for text and image processing. We present details of the dataset in the "Data collection and Description" section. Experimental results and discussion are presented in the next section. Finally we conclude and present future directions in the last section.

2. Artificial Intelligence Technologies for Disaster Response

For rapid crisis response, real-time insights are important for emergency responders (Imran *et al.*, 2014). To identify actionable and tactical information pieces from a growing stack of social media data and to inform decision making processes as early as possible, messages need to be processed as they arrive. Given the high volume of social media messages, we need to triage them by categorizing them into different actionable bins such as food, supplies, financial, logistics, *etc.* Manual analysis of these messages coming at high rates is not possible. Even if a large workforce could somehow be amassed to keep up with the high velocity of social media data, having humans do something that can be automated is inefficient and a misallocation of scarce resources. The problem gets even worse during prolonged crises, such as civil wars, or during the long recovery phase following major natural disasters.

Research on building automatic computational methods to overcome information overload issues has demonstrated significant advantages in many application domains (Nguyenet *et al.*, 2017; Nguyen *et al.*, 2016; Rudra *et al.*, 2016). For instance, machine learning techniques help recommend news articles to readers (Hwang *et al.*, 2015), and natural language processing techniques automate question-answering tasks both general purpose (Sasikumar 2014) as well as domain-specific *e.g.* helping health experts to answer HIV-related questions *via* SMS (Imran *et al.*, 2016). Despite advances in the field of artificial intelligence for standard web data sources, in which the techniques mainly deal with structured and errors free data, processing social media webdata is still challenging. Often messages on social media are brief, full of slang and contain mistakes such as misspellings and shortened forms.

2.1 Automatic Textual Content Processing

To overcome the above mentioned issues and shortcomings of the existing data processing techniques, we have developed an AI-based social media processing platform called artificial intelligence for digital response (AIDR). AIDR combines human-intelligence and machine learning techniques to process data. Specifically, AIDR uses supervised machine learning techniques to classify each message/tweet as it is posted on social media in real time. AIDR is specifically designed to scale up the abilities of human workers by intelligently removing noise form the data *e.g.* in the form of duplicate and irrelevant messages. At the on-set of a disaster event, humanitarian experts or members of affected communities start AIDR data collection. The data collection in AIDR can be configured either using event-specific keywords or hashtags; or by using geographical bounding boxes. One can define multiple bounding boxes to cover different disaster areas. In that case, geo-tagged tweets originating from the areas under surveillance are collected. However, we have observed that only 1 to 3 per cent of the tweets are geo-tagged. For this reason, AIDR also provides a combination of keywords and a geographical bounding box data collection strategy, in which case a user defines both keywords and bounding boxes to collect tweets either matching with one of the keywords or geo-tagged within one of the bounding boxes.

After setting the data collection, the user enables the machine classification process. This process involves defining a set of user defined categories of interest to process text messages. For instance, many humanitarian organizations are interested in learning about injured or dead people, infrastructure damage, urgent needs of affected people in terms of food, water, shelter, utilities, money *etc.* Depending on information needs, the user defines categories in AIDR, since AIDR uses supervised machine learning techniques, the next stage is to provide a handful of training examples for machine training. For this purpose, the user tags a few hundreds of tweets *i.e.* assign the appropriate category to tweets. AIDR learns new models using the training examples. New models are learned as more human tagged tweets become available. AIDR uses *Random Forest*, a well-known decision tree based learning scheme, to learn models. As for the features, uni-grams and bi-grams are used. A well-known feature selection technique called "Information Gain" is employed to select the top 1,000 features.

2.2 Automatic Multimedia Content Processing

In addition to textual messages, images available on social media could be useful for a number of humanitarian tasks. One of the core tasks, among others, that we are working on is automatic damage assessment from images. Disaster damage severity detection is an important situational awareness task through which humanitarian organizations estimate the scale of the disaster and accordingly plan response and reconstruction efforts. Critical infrastructure that is severely damaged gets high priority for repair or reconstruction. For this purpose, AIDR has been extended to process images on social media to determine the damage severity.

For image processing, AIDR employs state-of-the-art computer vision techniques. The image processing pipeline works in a similar way to the text processing pipeline, as explained earlier. For image processing, we also employ supervised machine learning techniques, which meanthat humans are in the loop to provide some sort of supervision to the machine. The supervision in this case is human tagged images. The data collection in this case is exactly same as explained earlier. The user specifies a set of keywords or hashtags or defines geographical bounding boxes for tweet collection. Each collected tweet is then checked to find if it contains any image link. In case of an image associated with the tweet, the image is downloaded from the Web. We have observed a large proportion of images posted during disaster events consist of duplicate content. Other than exact duplication, cropped, resized, recolored are other examples of duplicate in which the main scene stays similar to the original image. Furthermore, another big proportion of images on social media are irrelevant, showing cartoons, banners, celebrities, advertisements *etc.*

To tackle the issues of image duplication and relevancy, AIDR employs state-of-the-art techniques. For the de-duplication of images, we use *Perceptual Hashing Technique*. To determine, an image relevancy, we use *Convolutional Neural Networks*

(CNN). Specifically, we use VGG-16 trained on a large number of labeled images from the ImageNet repository (Jia *et al.*, 2009). We fine-tuned the same network using our labeled data after adapting the last layer.

After removing noise (*i.e.* duplicate and irrelevant) from the collected images, we perform damage assessment. We use three categories to represent damage or not damage as follows: we use the SEVERE damage category to represent non-useable buildings, non-drivable road or bridges; we use the MILD damage category to represent partially destroyed infrastructure; and the NO damage category to represent little to no damage. Given these categories, we employed human workers to tag images. Human-tagged images are then used to train CNN classifiers. Specifically, we adapt a transfer learning approach in which pre-trained models using the ImageNet data are fine-tuned using our own labeled data. For this purpose, we sampled 6,000 images from four disaster events, *viz.* the 2015 Nepal earthquake, the 2016 Ecuador earthquake, Typhoon Ruby in 2014 and Hurricane Matthew in 2016. The human-tagged images distribution is as follows: SEVERE = 1,765; MILD = 483; NO = 3,751.

2.3 Critical Information Extraction

Automatic approaches like AIDR help eliminate noise from the data. However, even after removing noisy content, the data may still be too big to be manually processed. Moreover, if the disaster event spans over weeks or months, this could create a filter failure issue. To overcome these issues, we have developed an interactive monitoring tool to help crisis responders and decision makers provide rapid access to the most critical and important situationalinformation during an on-going disaster event. For this purpose, we begin by extracting critical entities mentioned in tweets. These entities represent people, places, or organization names. We use the SNERT (Stanford Named Entity Recognition Tool) (Finkel *et al.*, 2005) for named entity extraction. Furthermore, we combine AIDR's classifiers output (*i.e.* classified tweets), use defined categories, timelines, and map components to help users quickly sift through thousands of tweets. **Figure 23.1** shows the interactive dashboard and its different visual components including timelines (top), machine classifiers (left), top critical entities (people, organizations, locations), and tweets view (bottom).

3. Hurricane Maria: Data Collection and Description

We collected twitter data during the 2017 Category 5 Hurricane Maria that made land fall around 20th Sept. 2017 and caused catastrophic damage in Puerto Rico, Dominica and the Dominican Republic. Around 129 people were killed by the hurricane in total,with 55 deaths in Puerto Rico and 57 in Dominica. The hurricane caused more than 51 billion US dollars in damage (Wikipedia 2017b).

To collect messages during the hurricane, we used the AIDR platform.AIDR uses the twitter streaming API and offers different data collection strategies such as collection by keywords or hashtags, by geographical bounding boxes, or by following specific twitter users. We used the keywords based data collection approach and formed a query consisting of event-specific keywords including

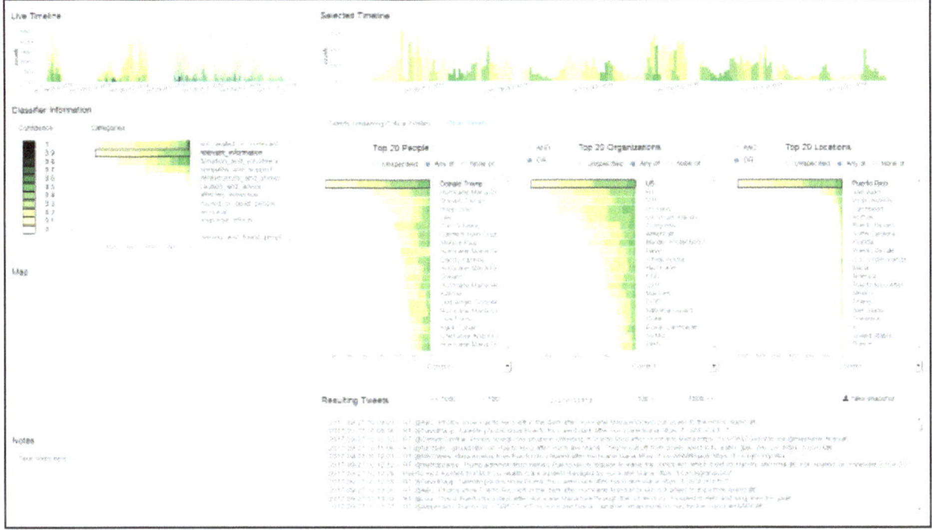

Figure 23.1: Interactive Dashboard to Quickly Find Critical and Useful Information during Disasters.

"Hurricane Maria", *"Maria storm"*, *"Maria cyclone"*, *"Tropical storm Maria"* and so on. As a result, in total, we collected around 2 million tweets from 20th Sept. 2017 to 06 Oct. 2017. **Figure 23.2** shows the distribution of tweets over these days. From the collected tweets, we found ~80k tweets with images. **Figure 23.3** shows a few images showing the destruction of the Hurricane Maria.

4. Results and Discussion

We employed AIDR's machine learning classifiers to understand the types

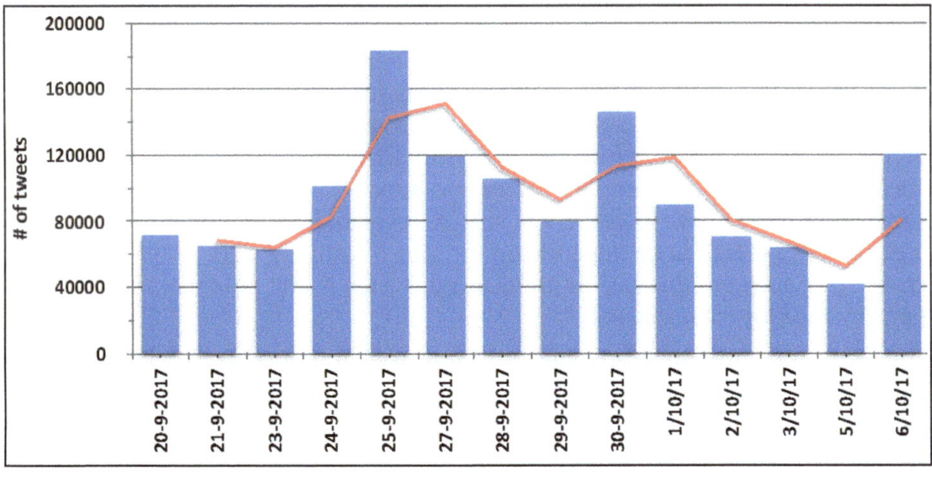

Figure 23.2: Hurricane Maria Number of Tweets per Day (The red trend line shows the moving average).

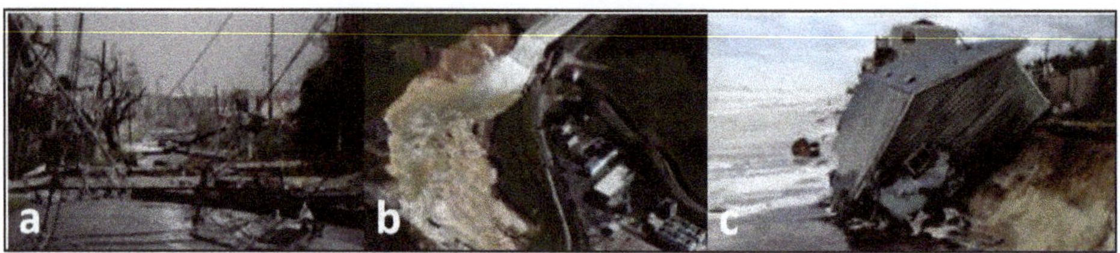

Figure 23.3: Sample Images Collected during Hurricane Maria.

of information present in the collected tweets. We trained a text classifier using ~30,000 human-labeled tweets, which were originallycollected from a number of past disaster events. Each tweet has been tagged by at least three human workers using the humanitarian categories shown in **Table 23.1**.

Table 23.1: Text Classifier Evaluation Results in Terms of Precision, Recall and F1

Category Name	*Precision*	*Recall*	*F1-score*
Affected individual	0.71	0.71	0.71
Caution and advice	0.61	0.64	0.62
Donation and volunteering	0.64	0.77	0.7
Infrastructure and utilities damage	0.65	0.65	0.65
Injured or dead people	0.82	0.87	0.84
Personal updates	0.66	0.62	0.64
Other relevant information	0.66	0.4	0.5
Sympathy and support	0.54	0.76	0.63
Not related or irrelevant	0.66	0.72	0.69

To train the classifier, we take 80 per cent of the human tagged data. For this purpose, we use the Random Forest learning scheme. The classifier evaluation is performed using a hold out test set consisting of 20 per cent of the labeled data. **Table 23.1** shows the evaluation results in terms of Precision, Recall, and F1. Overall, the classifier achieved an AUC score = 0.64, which is a reasonably acceptable performance. However, individually some classes have better scores such as injured or dead people F1=0.84 and affected individual F1= 0.71; whereas some other classes have lower scores, such as relevant information F1=0.50. Largely a classifier's performance depends on the size of training data available for each class. The dataset in this case is imbalanced, which is one potential reason of low performance for some classes.

We applied the trained classifier on the whole Hurricane Maria's dataset. **Figure 23.4** shows the prediction results in terms of the distribution of tweets into different classes. Consistent with our past observation while analyzing social media data during disasters and emergencies that the amount of irrelevant messages is always higher than the useful messages, in this case we also observed a similar pattern. The *"Not related or irrelevant"* category contains around 28 per cent of the whole data

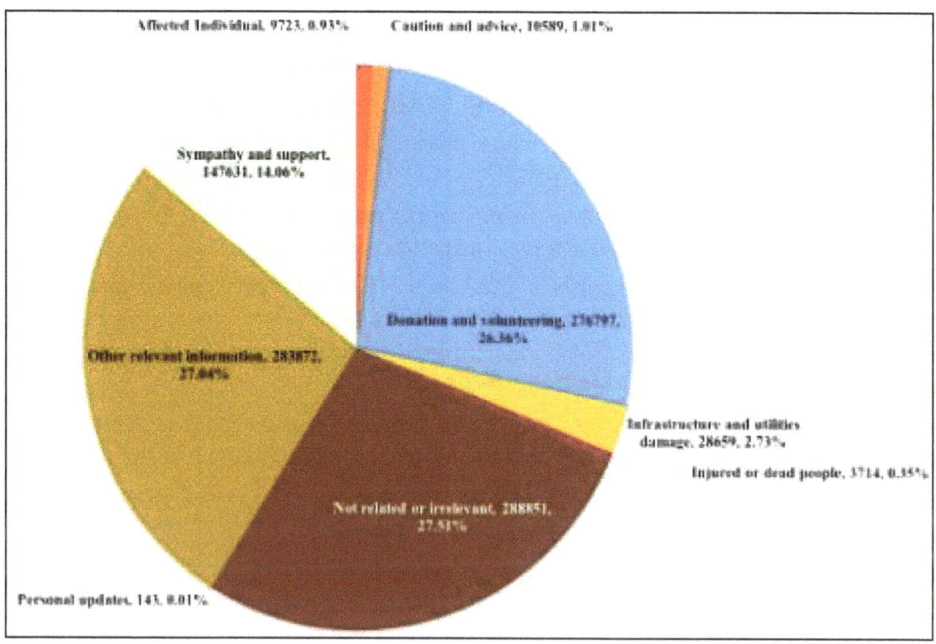

Figure 23.4: Tweets Distribution in different Categories by the Text Classifier.

in the Hurricane Maria case. The "Other relevant information" category emerged as the second largest category, which is also not surprising. The "Other relevant information" category contains messages that do not belong to any other informative class but these messages are important for humanitarian purposes. Research studies have found that this category contains from about 10 per cent to 30 per cent of the messages on the social media during disasters. For more information about why AIDR uses this category and why this category could be useful to learn about several small-scale unanticipated events, we refer to papers by Imran *et al.*, 2016.

In the remaining categories, the "*Donation* and *Volunteering*" and "*Sympathy* and *Support*" contain substantial amounts of messages, specifically 26 per cent and 14 per cent, respectively. Among other relatively small categories, the "*Infrastructure and Utilities Damage*" category contains around 3 per cent of the full data.

Next, we wanted to look into these individual categories and find out what kind of information these automatically categorized messages contain. However, we noticed that even after the automatic classification of tweets and the filtering-out of irrelevant ones, the remaining messages are still in the thousands. Manual analysis of these thousands of messages (*e.g.* ~27k donation-related messages) is a hectic and time-consuming task for crisis responders, especially during an on-going disaster event. To overcome this information-overload issue and to help crisis responders sift through enormous data as quickly as possible, we have developed an interactive dashboard technology that provides convenient ways to find critical information from big live data. We deployed the dashboard to analyze the Hurricane Maria's data. In below, we provide a few most critical tweets found during data analysis through the dashboard.

Donations and Volunteering Requests or Offers

- ✰ *RT @DorothyKidd1: Food, water, power, gasoline desperately needed in Puerto Rico https://t.co/z3tVTc7pAU*
- ✰ *Puerto Rico needs our help. Bring diapers, baby food, batteries, first aid supplies, feminine hygiene products to:... https://t.co/KWi6uCWMxz*
- ✰ *Our neighbors need our help! Click on the link on our page and donate what you can. Thank you! https://t.co/zyITNsMUHy*

Infrastructure and Utility Damage Reports

- ✰ *Bavaro, Dominican Republic A gift shop on Cofrecito beach is damaged after being hit by #HurricaneMaria Photograph:... https://t.co/51CmpqAtG2*
- ✰ *@Reuters Destroyed homes are seen from a Marine Corps Osprey surveying damage from #HurricaneMaria in St. Croix,... https://t.co/T7Xtc8vslP*
- ✰ *RT @youthmappers: There are 70k people downstream of this dam damaged by H.Mariahttps://t.co/cyki5UMZyq*
- ✰ *Hurricane Maria leaves Arecibo radio telescope damaged and dark https://t.co/b1G7EW9BEI via @theregister*
- ✰ *Hurricane Maria Damages Dominica's Main Hospital, Leaves 'War Zone' Conditions https://t.co/JuCd4XIkW8*
- ✰ *RT @NWSPittsburgh: Maria did a lot of damage to our Doppler Radar in San Juan, PR. Here is what a radar should look like....*
- ✰ *San Juan airport remains crippled by Hurricane Maria damage https://t.co/167dRw8OkX*
- ✰ *Maria's Devastation Damaged Hospitals Running Out of Supplies in U.S. Virgin Islands @weatherchannel https://t.co/BWfZmCRU1y*

Missing or Found People Reports

- ✰ *RT @Jasamsdestiny: Amy is missing – plz help find her. #Miami #southmiamiheights #palmettoestates #miamidadecounty #MiamiBeach...*
- ✰ *#hurricanemaria QUESTELL FAMILY MISSING 5 DAYS IN PONCE. PLEASE HELP FIND THEM*

During Hurricane Maria, AIDR's image processing pipeline was also activated to identify images that show infrastructure damage. In total, we found around 80k tweets containing images. However, ~75 per cent of these images were found to be duplicate. The remaining 25 per cent (~20k) of the images were automatically classified by the AIDR's damage assessment classifier. As stated above, we use three classes to represent the extent of damage in an image namely: SEVERE damage, MILD damage, and NO damage. From the predictions, we observe that almost 78 per cent of the images we collected on Twitter do not show any damage, mainly because these images depict hurricane paths, people and other info-graphics rather than built structures. In **Figure 23.5**, the first row shows some examples of such images. From the predictions, 10.9 per cent of images show MILD damage and 11.2 per cent of the images show SEVERE damage. **Figure 23.5** shows a few images automatically classified by our image classifier into the three damage classes.

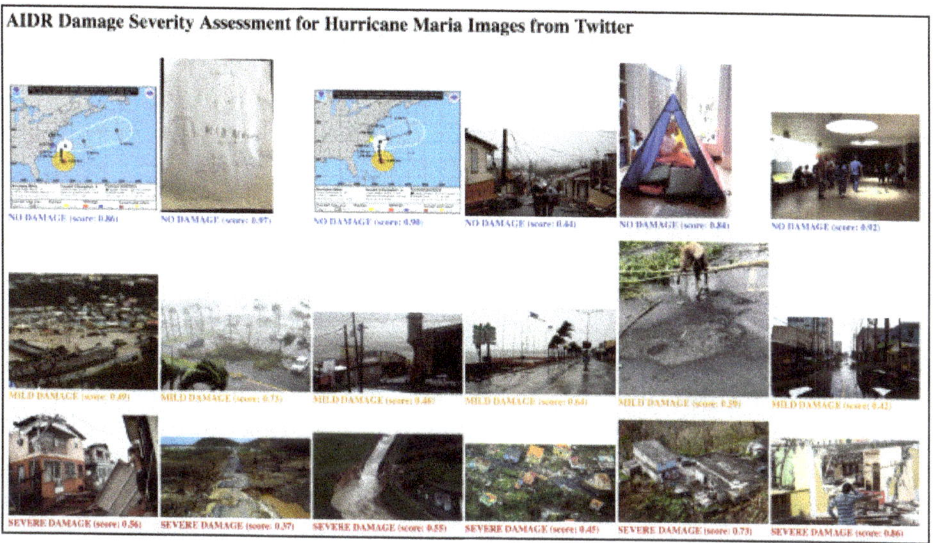

Figure 23.5: Damage Assessment Results Using Images Collected during Hurricane Maria.

By looking at the actual image content, we can extract more information about the devastation caused by the disaster than relying solely on the textual content provided by the users. The image processing pipeline could be useful for a number of humanitarian use cases. For instance, with the automatic damage assessment, instead of trying to look at all the images, humanitarian organizations and emergency responders can simply take a look at the images containing MILD or SEVERE damage to quickly understand sense of the level of destruction incurred by the disaster.

5. Conclusions

Humanitarian organizations rely on credible and timely information to act against natural disasters and emergencies. Social media has emerged as an alternative information source that contains information that directly comes from affected people. This information consists of a variety of useful reports such as reports of urgent needs of affected people and other situational information. However, real-time processing of social media data is a challenging task. In this work, we have presented a number of technologies that employ state-of-the-art artificial intelligence techniques to collect and process social media information in real-time to aid humanitarian organizations.

References

1. Alam, F., Imran, M. and Ofli, F. 2017. Image4Act: Online Social Media Image Processing for Disaster Response. In IEEE/ACM International Conference on Advances in Social Networks Analysis and Mining.
2. Aupetit, M. and Imran, M. 2017. Interactive monitoring of critical situational information on social media. In International Conference on Information Systems for Crisis Response And Management (ISCRAM).
3. Boersma, K., Fonio, C., Imran, M., and Meier, P. 2017. Big Data, Surveillance and Crisis Management. Routledge.
4. Castillo, C. 2016. Big crisis data: social media in disasters and time-critical situations. Cambridge University Press.
5. Daly, S. and Thom, J. A. 2016. Mining and classifying image posts on social media to analyse fires. 13th International Conference on Information Systems for Crisis Response and Management, (May), 1–14.
6. Finkel, J., Grenager, T. and Manning, C. 2005. Incorporating non-local information into information extraction systems by Gibbs sampling. Proceedings of the 43rd Annual Meeting on (June 2005), pp.363–370. https://doi.org/10.3115/1219840.1219885
7. Hughes, A. L. and Palen, L. 2009. Twitter adoption and use in mass convergence and emergency events. *International Journal of Emergency Management*, 6: 248.
8. Hwang, W.S., Park, J. and Kim, S.W. 2015. A method for recommending the latest news articles *via* MinHash and LSH. In Proceedings of the 9th International Conference on Ubiquitous Information Management and Communication - IMCOM '15 pp. 1–6.
9. Imran, M. and Castillo, C. 2015. Towards a data-driven approach to identify crisis-related topics in social media streams. In Proceedings of the 24th International Conference on World Wide Web Companion. pp. 1205–1210
10. Imran, M., Castillo, C., Diaz, F. and Vieweg, S. 2015. Processing social media messages in mass emergency: A survey. ACM Computing Surveys (CSUR), 47: 67.
11. Imran, M., Castillo, C., Lucas, J., Meier, P. and Rogstadius, J. 2014. Coordinating human and machine intelligence to classify microblog communications in crises. In 11th International Conference on Information Systems for Crisis Response and Management.
12. Imran, M., Castillo, C., Lucas, J., Meier, P. and Sarah, V. 2014. AIDR: Artificial Intelligence for Disaster Response. In Proceedings of the 23rd International World Wide Web Conference (WWW '14).
13. Imran, M., Chawla, S. and Castillo, C. 2016. A robust framework for classifying evolving document streams in an expert-machine-crowd setting. In Proceedings of the 16th International Conference on Data Mining (ICDM).

14. Imran, M., Meier, P., Castillo, C., Lesa, A. and Herranz, M.G. 2016. Enabling digital health by automatic classification of short messages. In Proc. of the ACM Digital Health 2016 Conference.
15. Jia D., Wei D., Socher, R., Li-Jia L., Kai Li. and Fei-Fei L. 2009. ImageNet: A large-scale hierarchical image database. In 2009 IEEE Conference on Computer Vision and Pattern Recognition. pp. 248–255
16. Nguyen, D.T., Al-Mannai, K., Joty, S.R., Sajjad, H., Imran, M. and Mitra, P. 2017. Robust classification of crisis-related data on social networks using convolutional neural networks. In ICWSM (pp. 632–635).
17. Nguyen, D.T., Alam, F., Ofli, F. and Imran, M. 2017. Automatic image filtering on social networks using deep learning and perceptual hashing during crises. *Info. Sys for Crisis Response and Management* (ISCRAM).
18. Nguyen, D.T., Joty, S., Imran, M., Sajjad, H. and Mitra, P. 2016. Applications of Online deep learning for crisis response using social media information. Social Web for Disaster Management (SWDM 2016) - Co-Located with CIKM 2016 - arXiv Preprint arXiv:1610.01030.
19. Nguyen, D.T., Ofli, F., Imran, M. and Mitra, P. 2017. Damage assessment from social media imagery data during disasters. In IEEE/ACM International Conference on Advances in Social Networks Analysis and Mining.
20. Ofli, F., Meier, P., Imran, M., Castillo, C., Tuia, D., Rey, N., *et al.*, 2016. Combining human computing and machine learning to make sense of big (Aerial) data for disaster response. *Big Data Journal*, 4: 47–59.
21. Rudra, K., Banerjee, S., Ganguly, N., Goyal, P., Imran, M. and Mitra, P. 2016. Summarizing situational tweets in crisis scenario. In Proceedings of the 27th ACM Conference on Hypertext and Social Media (pp. 137–147).
22. Sasikumar, U. 2014. A survey of natural language question answering system. *International Journal of Computer Applications*, 108: 975–8887.
23. Vieweg, S., Hughes, A. L., Starbird, K. and Palen, L. 2010. Microblogging during two natural hazards events/ : What Twitter May Contribute to Situational Awareness. CHI 2010: Crisis Informatics,
24. Whipkey, K. and Verity, A. 2015. Guidance for incorporating Big Data into humanitarian operations. Digital Humanitarian Network Document.

Web Resources Accessed

25. The Economist (2017). Weather-related disasters are increasing. https://www.economist.com/blogs/graphicdetail/2017/08/daily-chart-19, Published on 29th August, 2017.
26. Wikipedia (2017a). Hurricane Harvey. https://en.wikipedia.org/wiki/Hurricane_Harvey
27. Wikipedia (2005). Hurricane Katarina. https://en.wikipedia.org/wiki/Hurricane_Katarina
28. Wikipedia (2017b). Hurricane Maria. https://en.wikipedia.org/wiki/Hurricane_Maria

Chapter 24

A Review on Key Challenges Towards Risk Reduction Science and Innovation Plans

G. Clegg, N. Dias, D. Amaratunga and R. Haigh*

Global Disaster Resilience Centre,
University of Huddersfield, Queens Gate, Huddersfield, UK
**E-mail: n.dias@hud.ac.uk*

ABSTRACT

There are many obstacles and challenges to deliver a harmonized and coherent approach to the issue of risk reduction. Three of the key challenges which are often in the debate is the contrasting approaches traditionally used for Climate Change Adaptation (CCA) Vs. Disaster Risk Reduction (DRR), Science vs. legal/policy issues in risk reduction and national approaches to trans-boundary crises. Accordingly, it is now widely recognised the need to integrate CCA and DRR, match legal and policy background to absorb scientific innovations and the management of trans-boundary crisis among the nations. Based on findings of a desk-based literature review, conducted by a project called ESPREssO funded by the EU horizon 2020 programme, this paper reviews the key issues associated with the above mentioned key challenges which are integrating CCA and DRR, Matching scientific innovation with the legal and policy backgrounds and the trans-boundary crisis management. Findings reveal issues related to governance such as institutional barriers, funding issues, issues related to communication and issues related to the concept of risk which should be addressed in order to overcome these challenges.

Keywords: *Climate change adaptation, Disaster risk reduction, Science, Policy, Trans-boundary crises.*

1. Introduction

Natural disasters have major implications for society, the economy and the environment, resulting in widespread damage and loss of lives. As the global population increases, an increasing proportion is becoming exposed to natural hazards, increasing society's vulnerability. To highlight the scale, it is estimated that worldwide, exposure to natural hazards has doubled since 1975 (Pesaresi *et al.*, 2017) and approximately $7 trillion in damages has been incurred since 1990 (Amos, 2016). Of the total number of disasters, hydrological and meteorological disasters are most frequent (Guha-Sapir *et al.*, 2016). With climate change thought to be impacting the frequency, intensity and distribution of hydro-meteorological hazards, protecting society against these threats is becoming increasingly important.

Within this context, it is widely discussed that ***Climate Change Adaptation*** **(CCA)** and ***Disaster Risk Reduction*** **(DRR)** both look to increase resilience and reduce the vulnerability of society against these phenomena and there are several overlaps between the two practices (Mitchell and van Aalst, 2008). Although the convergence of CCA and DRR is recognised in theory, in reality, no clear strategy for integration exists, and the two operate independently. There are several existing challenges that must be overcome to help foster integration.

Also, the role of science in effective decision making for climate change and disaster policy is paramount (Basher, 2013), but, in the current context, barriers exist which prevent science being used optimally in policy making.

Further to this, The Sendai Framework for Disaster Risk Reduction (SFDRR), a global framework to reduce disaster risk and loss of lives (Basher, 2013; UNISDR, 2015) highlights the management of trans-boundary disasters as one of the major unsolved problems. A trans-boundary crisis is a threat to several different geographical or policy domains. Trans-boundary crises pose a challenge to management as they do not respect the boarders that organise administrative and political response capacities (Boin *et al.*, 2017).

Taking these three concerns on board, the **ESPREssO** project (*Enhancing Synergies for Disaster Prevention in the European Union*), which is funded by the European Commission Horizon 2020 research and innovation programme, to which this paper contributes, looks to address the following three challenges:

1. Identify ways to create more coherent national and European approaches on Disaster Risk Reduction, Climate Change Adaptation and resilience strengthening.
2. To bridge the gap between policy and science at local and national levels in six European countries in order to enhance risk management capabilities.
3. To address the issue of trans-boundary crises and to achieve more efficient management.

The three ESPREssO challenges provide the basis for this report, in which the gaps, challenges and inconsistencies within the challenges will be explored in order to contribute to ESPREssO's overarching aim, to contribute a new strategic vision to approach natural risk reduction and climate change adaptation, thereby opening new frontiers for research and policy making.

2. Methodology

As described in section 01 this paper is based on the ESPREssO project. Accordingly, as a part of the project a scientific literature review has been completed on the three ESPREssO Challenges and this discussion paper is purely based on a desk-based narrative literature study. This thorough literature review conducted to evaluate the state of the art in the subject area consists, findings from a range of peer-reviewed journals, books, conference proceeding, high-level reports from global organisations and so on.

3. Discussions

This section describes the key gaps and issues related to the three ESPREssO challenges and the consequences of it.

3.1 Integration of CCA and DRR

The synergies between CCA and DRR are becoming increasingly recognised (Shaw *et al.*, 2010) and a systematic linkage between the two has clear benefits in achieving common goals and advancing sustainable development. However, there are several challenges which create barriers to their successful integration resulting in the two presently operating as separate practices.

In many countries around the world, CCA and DRR policies are managed at the national level by government institutions. The government structure is commonly sectoral, with departments working independently. As a result, CCA and DRR activities have developed separately and operate in parallel. CCA commonly falls within the domain of the environment ministry, while DRR is the focus of infrastructure development ministries. For example, in the United Kingdom, disaster management is governed by the Civil Contingencies Secretariat and climate change activities by the Department for Environment, Food and Rural Affairs (DEFRA). Having developed separately, the two departments have different cultures and perceptions, with CCA having developed in the scientific community and DRR within the humanitarian (Mitchell and van Aalst, 2008). This institutional arrangement prevents the CCA and DRR communities working together, firstly, through hindering communication and the transfer of information (Schipper and Pelling, 2006). Limited communication between the two communities means agendas and frameworks are developed separately and could lead to the duplication of work and inefficient use of resources.

There is low political will for DRR compared to CCA which is present in both developed and developing countries (Dupuis, 2011). This may be influenced by the public greater engagement with climate change issues and presents a challenge to integration, as DRR and CCA are not seen to have equal political value.

Due to greater political will for CCA, funding is greater than that for DRR. Due to the segmented institutional structure, funding for CCA and DRR comes from different streams. Funding organisations may be disinclined to integrate CCA and DRR funding due to potentially greater expense of doing so (Mitchell and van Aalst, 2008). This results in funding being another common barrier to CCA and DRR integration (Biesbroek *et al.*, 2010).

In summary, existing institutional arrangements are a primary barrier to the integration of CCA and DRR as the divided structure results in limited communication and unequal funding, enforcing the gap between CCA and DRR communities.

3.2 Science and Policy

Successful risk management is dependent on the production of new knowledge and innovative technologies and for this information to be transferred to policy makers and practitioners in order to support the decision-making process. However, Biesbroek *et al.* (2010) note that science and policy do not automatically harmonise, with large quantities of new knowledge being produced by scientists that does not find its way into the policy domain.

In academia, complex language is common place, representing the specialised nature of the topic being studied. However, decision makers are often non-experts and may find the complex language difficult to understand. In general, there is a lack of universal terminology for CCA and DRR, resulting ambiguity and confusion when those from different organisations try to communicate (Kelman *et al.*, 2017). In this way, communication can be identified as preventing integration of science into policy making.

There are differences in the procedures of scientists and policy makers. For example, policy makers are often required to make decisions quickly, while researchers require time to discern a conclusion and are disinclined to give a premature statement (Woo and Marzocchi, 2014). Scientists often operate in terms of probability, while decision makers follow Boolean logic (Woo and Marzocchi, 2012). These operational differences result in poor alignment of the two communities. Better understanding of each other's methods would help to achieve better coordination.

Thus, there is a requirement for scientists to produce research that is applicable to, and use able for, policy making. However, it is not only science on which policy decisions are made, there are also political factors that come into consideration and may compete with scientific evidence. As follows, the challenge is for scientists to produce policy relevant research but to not become politicised, in order to maintain legitimacy and credibility (Biesbroek *et al.*, 2010). In a study by (Lorenzoni *et al.*, 2007) university scientists were found to be perceived as the most trustworthy source of information on climate change. Therefore, academics should play a key role in the dissemination of information. As a result, the challenges are thus; how can scientists produce policy relevant research and communicate this in an easy to understand way, while maintaining legitimacy?

Presently there is a limited arena for exchanges between scientists and policy makers to take place. A suggested mechanism for creating a two-way exchange between scientists and policy makers is bridging organisations to facilitate the exchange of information between information producers and information users

(Kirchhoff *et al.*, 2015). But who should take responsibility for the set up and management of such organisations remains in question.

3.3 Trans-boundary Crisis Management

The SFDRR (2015) identifies the management of trans-boundary crises as one of the main challenges pertaining to disaster resilience at the global level. Many disasters have impacts that cross national boundaries, for example, the 2003 European heatwave and drought which impacted a large proportion of central Europe. To achieve an effective response to trans-boundary disasters, multiple organisations must coordinate together (Boin and Lodge, 2016). In practice however, the successful management of trans-boundary crisis complex and coordination commonly occurs as a result of a disaster, in a reactive, rather than proactive manner (Lebel *et al.*, 2010) while, coordination remains a significant challenge.

A clear emergency operations plan is suggested as a requirement for successful management of trans-boundary crises (Edwards, 2009). But as (Boin and McConnell, 2007) point out, such plans often bear little relevance to the challenges that emerge during a crisis, as due to their very nature, crises are unprecedented and have unforeseen consequences.

But it is unclear who should take responsibility for providing the coordinated effort and producing such operations plans. The international community has a role to play in coordination of nations, but the effectiveness of response is generally the responsibility of local actors (IFRC, 2015), therefore there is also the added challenge of vertical integration, from the national to the local level, as well as horizontally between nations.

Further to this, mismatch between nations creates further complexity. Countries are exposed to different hazards and have different needs during a disaster and the capacity to respond to disasters varies greatly between countries (Siciliano and Wukich, 2017). In addition, different stakeholders prioritise and understand problems and solutions differently and during a wide-area disaster there are many stakeholders involved (Frankish *et al.*, 2012). Therefore, the problem is not a simple case of coordination but requires careful consideration of the needs and capabilities of different countries.

4. Summary

With regards to the challenge integrating CCA and DRR, this literature review highlighted several research gaps that need to be addressed to better understand the reasons for the separation between CCA and DRR. While the interdependencies between CCA and DRR are evident (Becker, 2009), the practitioners' and researchers' communities dealing with these issues still seem to remain isolated from each other (Gaillard, 2010). On one side, it is still not clear how to achieve CCA outcomes through improved DRR policies, planning and risk management. On the other side, it is not clear what added value CCA measures bring to existing DRR tools and plans.

In regards to the second challenge of science *vs.* legal and policies, it is identified that communication among scientists, decision-makers, and citizens seems to be a key aspect to be improved. One way of doing this could be the need for a larger inclusion of social sciences in disaster risk management. In fact, social sciences can provide the necessary perspective to warn and inform population and different sectors of society more efficiently. Further, understanding of risk across different spheres in society, and how to remedy this, is also an issue that should be dealt. Education seems to be crucial for this process. Public risk understanding is a key objective to pursue in order to enhance citizens engagement in disaster risk management.

Accordingly, this thirty months project called ESPREssO seeks to find out possible solutions for these challenges and finally to develop a new strategic vision to approach natural risk reduction and climate change adaptation, thereby opening new frontiers for research and policymaking.

Acknowledgements

This project has received funding from the European Union's Horizon 2020 research and innovation programme under grant agreement 700342. This publication reflects the views only of the author, and the Commission cannot be held responsible for any use which may be made of this information contained therein.

References

1. Amos J. 2016. Economic losses from natural disasters counted. *BBC News*, 18th April 2016.
2. Basher R. 2013. Science and Technology for Disaster Risk Reduction: A Review of Application and Coordination Needs. Geneva: Secretariat to the UN International Strategy for Disaster Reduction (UNISDR).
3. Biesbroek G.R., Swart R.J., Carter T.R., Cowan C., Henrichs T., Mela H., Morecroft M.D. and Rey D. 2010. Europe adapts to climate change: Comparing National Adaptation Strategies. *Global Environmental Change* 20: 440-450
4. Boin A., Hart P.T., Stern E. and Sundelius B. 2017. The politics of crisis management: public leadership under pressure. Cambridge University Press, Cambridge
5. Boin A. and Lodge M. 2016. Designing resilient institutions for trans-boundary crisis management: A time for public administration. *Public Administration* 94: 289-298
6. Boin A. and Mcconnell A. 2007. Preparing for critical infrastructure. *Journal of Contingencies and Crisis Management* 15: 50-59.
7. Dupuis J. 2011. Political barriers to climate change adaptation development and society: Climate Change, Politics.

   https://ourworld.unu.edu/en/political-barriers-to-climate-change-adaptation: *Swiss Graduate School of Public Administration.*
8. Edwards F.L. 2009. Effective disaster response in cross border events. *Journal of Contingencies and Crisis Management* 17: 255-265

9. Frankish J.S., Roberts R.G., Coad A., Spears T.C. and Storey D.J. 2012. Do entrepreneurs really learn? Or do they just tell us that they do? *Industrial and Corporate Change* 22: 73-106
10. Guha-Sapir D., Hoyois P. and Below R. 2016. Annual Disaster Statistical Review 2015
11. IFRC 2015. World Disasters Report: Focus on Local Actors, the key to humanitarian effectiveness. In: HAMZA, M. (Eds.). Geneva, Switzerland: International Federation of Red Cross and Red Crescent Societies.
12. Kelman I., Mercer J. and Gaillard J.C. 2017. The Routledge Handbook of Disaster Risk Reduction Including Climate Change Adaptation, Oxon., Routledge
13. Kirchhoff C.J., Lemos M.C. and Kalafatis S. 2015. Narrowing the gap between climate science and adaptation action: the role of boundary chains. *Climate Risk Management* 9: 1-5.
14. Lebel L., Sinh B.T. and Nikitina E. 2010. Adaptive Governance of Risks: Climate, Water and Disasters. In: Shaw, R., Pulhin, J. M. and Pereira, J. J. (eds.) Climate Change and Disaster Risk Reduction:issues and Challenges. Bingley, UK: Emerald.
15. Lorenzoni I., Nicholson-Cole S. and Whitmarsh L. 2007. Barriers perceived to engaging with climate chnage among the UK Public and their Policy Implications. *Global Environmental Change* 17: 445-459
16. Mitchell T. and Van Aalst M. 2008. Convergence of Disaster Risk Reduction and Climate Change Adaptation A Review for DFID 1–22.
17. Pesaresi M., Ehrlich D., Kemper T., Siragusa A., Florczyk A.J., Freire S. And Corbane C. 2017. JRC Science for Policy Report: Atlas of the Human Planet: *Global Exposure to Natural Hazards*. Luxembourg
18. Schipper L. and Pelling M. 2006. Disaster risk, climate change and international development: scope for, and challenges to, integration. *Disasters* 30: 19-38
19. Shaw R., Pulhin J. M. and Pereira J.J. 2010. Climate change adaptation and disaster risk reduction: overview of issues and challenges. Climate Change Adaptation and Disaster Risk Reduction: Issues and Challenges. *Community, Environment and Disaster Risk Management* 4: 1-19
20. Siciliano M. and Wukich C. 2017. Network formation during disasters: exploring micro-level inter organizational processes and the role of national capacity. *Internation Journal of Public dminisatration* 40: 490-503
21. Woo G. and Marzocchi W. 2012. Previsione operativa dei terremoti e decisioni. *Amb. Risc. Commun* 4: 21-25
22. Woo G. and Marzocchi W. 2014. Operational earthquake forecasting and decision-making. *Early Warning for Geological Disasters*. Berlin: Springer.
23. Amaratunga D., Haigh R., Dias, Hemachandra K., Scolobig A. and Marx S. 2016. Espresso Project, Internal Report on Challenge 01- Climate Change Adaptation

and Disaster Risk Reduction. HUD, The UK, ETHZ, Switzerland and DKKV, Germany

24. Lauta K., Raju E., Kielberg M.F. and DI Ruocco A. 2016 ESPREssO Project, Internal report on Challenge "Science vs Legal/policy issues in DRR". UCPH, Denmark and AMRA, Italy
25. Parolai S., Fleming K., Ettinger S. And Baills A. 2016. ESPREssO project, Internal report on Challenge 3 "National regulations for the preparation to trans-boundary crises". GFZ, Germany and BRGM, France

Colombo Resolution on

Mitigation of Impacts of Human Hazards due to Extreme Natural Events Five Year Road-Map on the Adoption and Development of Scientific and Technological Advancements

RECOGNIZING THAT the natural extreme events prevalent in most countries lead to the loss of thousands of lives and billions of dollars in the form of property and environmental damage, the mitigation of which requires innovation of new technological advances and immediate intervention, improvisation and adoption of existing technologies through collective efforts, and knowledge and resource sharing among the regional countries experiencing similar types of disasters;

RECALLING THE SEVERITY of several incidents, for example, recent floods, landslides, droughts and earthquakes in India, Indonesia, Iran, Iraq, Jordan, Malaysia, Mexico, Nepal, Pakistan, Sierra Leone, Sri Lanka, and Thailand;

IDENTIFYING the extreme natural events that adversely affect most of the countries in the world which include Cyclones and Heavy Precipitation, Prolonged Droughts, Earthquakes and Volcanic Eruptions, Thunderstorms and Lightning, Hailstorms, Dust Storms, Cloudbursts and GLOF, Wind Gusts, Gales and Tornadoes, Heat and Cold waves, Acid Rains and Tsunami;

REALIZING THAT each of the above phenomena may give rise to one or more of the adverse effects that lead to loss of human lives and sufferings due to Flash Floods, Landslides, Impacts on Forestry and Vegetation, Impacts on Water Resources, Effects on Man-Made Structures, Health and Sanitation Issues, Impacts on Fisheries Industry, Impacts on Agriculture, Deterioration of Surface Topography and Man-Made Structures;

FURTHER ACKNOWLEDGING the scientific and technical advancements available at present for adoption as measures of mitigating the adverse effects due to extreme natural events including Advanced Forecasting, Now casting and Early Warning Systems; Qualitative and Quantitative Prediction of Adverse Effects; Space-Borne, Air-Borne and Terrestrial Remote Sensing; Mapping, Zoning and Simulation by GIS; Topographic Studies by Ground-Based and Airborne Platforms; Monitoring and Surveillance Systems; IT-Based Coordination Networks for Disaster Risk Reduction; Disaster-Resilient Housing and Sheltering; Earth Stabilization Techniques; Disaster-Resilient Electrical Systems; Urban Planning for Disaster-Resilient Living; Protection Systems and Safety Structures; Smart Evacuation Techniques; Chemical or Physical Treatment for Safety and Recovery; Sociological Measures for Disaster Mitigation; and Public Communication on S&T and Buffer Zoning to prevent Public Activities in Potentially Hazardous Areas.

WE, THE DELEGATES OF THE INTERNATIONAL ROUNDTABLE on the Impact of Extreme Natural Events: Science and Technology for Mitigation (IRENE), jointly organized by the Centre for Science and Technology of the Non-Aligned and Other Developing Countries (NAM S&T Centre), New Delhi, India, the National Science and Technology Commission (NASTEC), Sri Lanka and the Research Centre – Technology for Disaster Prevention (RC-TDP), South Eastern University of Sri Lanka, held in Colombo, Sri Lanka from 13th to 15th December 2017, representing 18 countries including Egypt, India, Indonesia, Iran, Iraq, Malaysia, Mauritius, Myanmar, Nepal, Pakistan, Palestine, Qatar, South Africa, Sri Lanka, Togo, the UK, Vietnam and Zambia;

AFTER EXTENSIVE DELIBERATION on various facets of transferring science and technology for mitigating hazards due to extreme natural events unanimously resolve to propose the following Five-Year Road-Map with implicit understanding that while the main stakeholder of this task is undoubtedly the Government, the success depends on the involvement and devotion of a large spectrum of stakeholders that include non-governmental organizations, R&D sector and experts in the field, academia, social workers, local authorities and many elements of each societal layer:

- ✰ Each country or region establishes an Activity Centre(s) (research, awareness or technical/consultancy services) affiliated to the Government, NGO, academia or standalone to address the relevant issues;
- ✰ An international or regional hub(s), possibly selected among the above Centres, is recognized to facilitate inter-institutional communication, networking, joint activities and regional collaborations, knowledge sharing and resource optimization;
- ✰ Each Centre acts as a knowledge and networking hub for the communities in the respective country or region;
- ✰ Each Centre identifies the most likely extreme natural events and their consequential adverse effects in the respective country or regio nd

educate the concerned stakeholders regarding the risks and threats in pre- and post- disaster scenario;

- Each Centre formulates, or adopts from the knowledge bases of other Centres, the latest technological advancements available to mitigate identified potential adverse effects due to natural extreme events in their respective countries or regions;
- The Centres individually or collectively approach the R&D sector and academia to encourage the relevant parties in inventing, innovating, improvising and adopting appropriate technologies to address the disaster related issues;
- The Centres individually or collectively develop a dialogue between the researchers and funding agencies to support the projects in the concerned areas;
- Each Centre builds up a close and cordial relationship with the government agencies to make them aware of the needs of the country or region and draw their attention in developing appropriate policies and guidelines;
- Each Centre works to seek funding from state and private sectors for research and development, awareness promotion, communication activities and international events to educate the local, regional and international communities regarding the technological and scientific solutions available for disaster risk reduction.

The participants of the Roundtable warmly applauded the proposal of Prof. B C Panda of Indira Gandhi Institute of Technology, Odisha, India to explore in-principle the possibility of establishing a dedicated NAM S&T Centre for Lightning Research, Detection and Protection (CLRDP) with the involvement of the State and Central Governments of India.

Thus, adopted this day, the 14th of December 2017, at Colombo, Sri Lanka.

Index

Q

R

S

T

U

V

W

Z

www.ingramcontent.com/pod-product-compliance
Ingram Content Group UK Ltd.
Pitfield, Milton Keynes, MK11 3LW, UK
UKHW020141300726
14059UKWH00007B/48